Simula SpringerBriefs on Computing

Volume 16

About this series

In 2016, Springer and Simula launched the book series *Simula SpringerBriefs on Computing*, which aims to provide introductions to selected research topics in computing. The series provides compact introductions for students and researchers entering a new field, brief disciplinary overviews of the state-of-the-art of select fields, and raises essential critical questions and open challenges in the field of computing. Published by SpringerOpen, all *Simula SpringerBriefs on Computing* are open access, allowing for faster sharing and wider dissemination of knowledge.

Simula Research Laboratory is a leading Norwegian research organization which specializes in computing. Going forward, the book series will provide introductory volumes on the main topics within Simula's expertise, including communications technology, software engineering and scientific computing.

By publishing the *Simula SpringerBriefs on Computing*, Simula Research Laboratory acts on its mandate of emphasizing research education. Books in this series are published by invitation from one of the series editors. Authors interested in publishing in the series are encouraged to contact any member of the editorial board.

Yan Zhang

Digital Twin

Architectures, Networks, and Applications

Yan Zhang
University of Oslo
Oslo, Norway

ISSN 2512-1677   ISSN 2512-1685 (electronic)
Simula SpringerBriefs on Computing
ISBN 978-3-031-51818-8   ISBN 978-3-031-51819-5 (eBook)
https://doi.org/10.1007/978-3-031-51819-5

This Springer imprint is published by the registered company Springer Nature Switzerland AG
The registered company address is: Gewerbestrasse 11, 6330 Cham, Switzerland

Paper in this product is recyclable.

Series Foreword

Dear reader,

Scientific research is increasingly interdisciplinary, and both students and experienced researchers often face the need to learn the foundations, tools, and methods of a new research field. This process can be quite demanding, and typically involves extensive literature searches and reading dozens of scientific papers in which the notation and style of presentation varies considerably. Since the establishment of this series in 2016 by founding editor-in-chief Aslak Tveito, the briefs in this series have aimed to ease the process by introducing and explaining important concepts and theories in a relatively narrow field, and to outline open research challenges and pose critical questions on the fundamentals of that field. The goal is to provide the necessary understanding and background knowledge and to motivate further studies of the relevant scientific literature. A typical brief in this series should be around 100 pages and should be well suited as material for a research seminar in a well-defined and limited area of computing.

We publish all items in this series under the SpringerOpen framework, as this allows authors to use the series to publish an initial version of their manuscript that could subsequently evolve into a full-scale book on a broader theme. Since the briefs are freely available online, the authors do not receive any direct income from the sales; however, remuneration is provided for every completed manuscript. Briefs are written on the basis of an invitation from a member of the editorial board. Suggestions for possible topics are most welcome and can be sent to sundnes@simula.no.

March 2023

Dr. Joakim Sundnes
Editor-in-Chief
Simula Research Laboratory
sundnes@simula.no

Dr. Martin Peters
Executive Editor Mathematics
Springer Heidelberg, Germany
martin.peters@springer.com

Series Editor for this Volume

Olav Lysne, Simula Metropolitan Center for Digital Engineering, Oslo, Norway

Preface

Digital Twin: Architectures, Networks, and Applications offers comprehensive, self-contained knowledge on Digital Twin (DT), which is a highly promising technology for achieving digital intelligence and digitally transformed society. DT is a key technology to connect physical systems and digital spaces. A digital twin is defined as the real-time digital replica of a real-world physical object. Digital twin in the digital space is able to monitor, design, analyze, optimize and predict physical systems. The bi-directional interaction between physical spaces and digital spaces brings many advantages, including low maintenance cost, reduced security risk, and substantially increased Quality-of-Service. Digital twin can also create unprecedented applications and services, ranging from Extended Reality (XR), immersive multimedia to remote medical care, autonomous driving, Web 3.0 and Metaverse.

The objectives of this book are to provide the basic concepts of DT, to explore the promising applications of DT integrated with emerging technologies, and to give insights into the possible future directions of DT. For easy understanding, this book also presents several use cases for DT models and applications in different scenarios. This book has the following salient features:

- Provides a comprehensive reference on state-of-the-art technologies for digital twin
- Covers basic concepts, techniques, research topics and future directions
- Contains illustrative figures that enable easy understanding of digital twin
- Allows complete cross-referencing owing to the broad coverage on digital twin
- Identifies the unique challenges for efficiently improving the performance of digital twin networks

This book allows an easy cross-reference owing to the broad coverage on both the principle and applications of DT. It provides a comprehensive technical guide covering basic concepts, innovative techniques, fundamental research challenges, recent advances and future directions on DT. The book starts with the basic concepts, models, and network architectures of DT. Then, we present the new opportunities when DT meets edge computing, Blockchain and Artificial Intelligence, and distributed machine learning (e.g., federated learning, multi-agent deep reinforcement learning).

In the last part, we present a wide application of DT as an enabling technology for 6G networks, Aerial-Ground Networks, and Unmanned Aerial Vehicles (UAVs). We also identify the future direction of DT in Reconfigurable Intelligent Surface (RIS) and Internet of Vehicles.

The primary audience includes senior undergraduates, postgraduates, educators, scientists, researchers, engineers, innovators and research strategists. This book is mainly designed for academics and researchers from both academia and industry who are working in the field of telecommunications, computer science and engineering, and digitalization. Students majoring in computer science, electronics, and communications will also benefit from this book.

October, 2023   *Yan Zhang*

Acknowledgements

This early morning when I am about to write the preface and acknowledgement, I came to realize that today is a very special day in Norway, "Grunnlovsdagen" means National day in Norwegian. Many people near Oslo travel to the city center and celebrate the important day with happiness, mostly importantly without using mask. In February 2022, Norway decided to lift all COVID-related restrictions. I would like to take this opportunity to wish you all healthy, safe, and joyful every day.

I started to think of the research challenges related to Digital Twin from April 2020 when we had to stay at home due to COVID-19. After fruitful discussions with collaborators and students, we came to understand the potential as well as the unique challenges of DT. I am very proud that we are now leading this research field of Digital Twin, demonstrated by several landmark "Hot Papers" and "Highly Cited Papers" in this field. We invented new terms including Digital Twin Networks, Digital Twin Edge Networks, and Wireless Digital Twin Networks, which are quickly recognized and followed by the scientific community and industry. We also identified several fundamental and unique challenges related to Digital Twin, e.g., edge association, which reflects the essential physical-digital interaction. My gratefulness should go to all of the excellent students and research collaborators Yulong Lu, Yueyue Dai, Wen Sun, Sabita Maharjan, Li Jiang, and Ke Zhang and many others. I appreciate all their contributions of time, discussions and ideas to study this exciting research field digital twin as well as make this book possible.

Special thanks go to Department of Informatics (IFI) at University of Oslo (UiO) where I work full-time from 2016, which is one of the most important periods of my career development. As a world-leading university over the world, UiO offers the chances to collaborate with the established professors, and the smart and hard-working students in IFI. UiO strongly encourages an open, collaborative and student-oriented environment. I am so lucky to have freedom to vision the future research direction and then establish our research reputation in the field of edge intelligence, blockchain, energy informatics and recently digital twin.

Special thanks go to Simula Research Laboratory and Simula Metropolitan Center for Digital Engineering where I worked full-time during 2006-2016 and part-time after 2016 until now. The most important scientific contributions for elevating me to IEEE Fellow have been carried out at Simula Research Laboratory. As an adjunct chief research scientist, I still receive strong support from Simula Research Laboratory, for which I am always much appreciated.

I am very grateful for the staffs at Springer for their support, patience and professionalism since the beginning until the final stage. All of them are extremely professional and cooperative. Last but not least, I want to give my deep thanks to my families and friends for their constant encouragement, patience and understanding throughout this project.

Yan Zhang 17.05.2022

Contents

Acronyms

6G	Sixth generation mobile networks
AI	Artificial intelligence
AR	Augmented reality
BS	Base stations
D2D	Device to device
DDPG	Deep deterministic policy gradient
DITEN	Digital twin edge networks
DPoS	Delegated proof of stake
DRL	Deep reinforcement learning
DT	Digital twin
DTN	Digital twin networks
EOS	Enterprise Operation System
IIoT	Industrial Internet of things
IoT	Internet of things
IoV	Internet of vehicles
ITS	Intelligent transportation system
LSTM	Long short-term memory
MBS	Macro base station
MIMO	Massive multiple-input-multiple-output
NFV	Network function virtualization
PoW	Proof-of-work
PLM	Product life-cycle management
RIS	Reconfigurable intelligence surface
RSU	Roadside unit
SDN	Software defined networking
TDMA	Time division medium access
UAV	Unmanned aerial vehicles
V2I	Vehicle-to-infrastructure
V2P	Vehicle-to-pedestrian
V2R	Vehicle-to-RSU
V2S	Vehicle-to-sensor

V2V	Vehicle-to-vehicle
VEC	Vehicular edge computing
VR	Virtual reality
XR	Extended reality

Chapter 1
Introduction

Abstract This chapter first gives an overview of digital twin, including the development timeline and possible application areas. The features of digital twin are summarized in detail. Then, the concepts, fundamentals, and visions of digital twin are presented.

1.1 Overview of Digital Twin

In recent years, digital twin has emerged as one of the most promising enabling technologies for sixth-generation (6G) mobile networks. Both academic and industry have shown increasing interest in unlocking the potential applications of digital twin in a range of areas, including smart cities, intelligent transportation, healthcare, energy, and Industrial Internet of Things (IIoT).

The overall development timeline of digital twin is shown in Fig. 1.1. The term *digital twin* was first introduced in 2002 by Michael Grieves in a presentation about product life cycle management. Thereafter, Framling et al. [1] proposed an agent-based architecture that maintains a corresponding virtual counterpart or agent for each product item with a faithful view of the product status and information [2]. These works advanced the initial exploration of digital twins from a conceptual point of view. Before 2010, NASA put the digital twin concept into practical application by developing two identical space vehicles for the Apollo project that could simulate and reflect the flight status in training. Since then, the idea of digital twins has been explored in areas such as aircraft maintenance and air force product management [3]. In 2017, Grieves gave the formal definition of digital twin in a digital twin white paper [4]. This white paper presented the basic digital twin model, which consists of physical objects, virtual objects, and a data link between physical space and virtual space. Recently, digital twin has been widely investigated in the areas of the IIoT and manufacturing for applications such as predictive diagnosis, production planning, and performance optimization [5]. Gartner listed digital twin as one of the top 10 strategic technologies in 2017, predicting that millions of things would have digital

Y. Zhang, *Digital Twin*, Simula SpringerBriefs on Computing 16,
https://doi.org/10.1007/978-3-031-51819-5_1

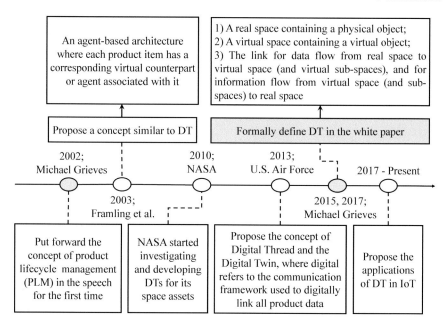

Fig. 1.1 The development timeline of digital twin

twin representations within three to five years. Gartner also listed digital twin as one of the top 10 strategic technologies in the next two years [6, 7], which shows industry's great confidence in digital twin technology. Gartner's top 10 strategic technologies for 2018 are shown in Fig. 1.2, with digital twin listed at the peak of inflated expectations.

More recently, digital twin has attracted a great deal of attention and has been widely explored in aviation, healthcare, smart cities, intelligent transportation, urban intelligence, and future wireless communications. The digital twin fulfils the role of collecting the real-time and historical running status of physical objects and making corresponding predictions and optimized decisions to improve the running performance of physical systems. In the field of aviation, digital twin has been used for aircraft maintenance, structuring, and risk prediction. For example, the authors in [8] proposed leveraging digital twin to model aircraft wings in the detection of fatigue cracks. In healthcare, with the assistance of wearable sensors and Internet of Things (IoT) devices, digital twin can be used to collect detailed physiological status and medication data about patients, which can help to monitor their medical condition and provide them with advanced medical care. In intelligent transportation systems, digital twins can help manage traffic, plan driving paths, and maintain transportation facilities. Real-time traffic conditions and the status of transportation facilities can be mirrored and analysed by digital twins in these transportation systems. Moreover, although the application of digital twins to scenarios of wireless communications is still in its infancy, several works can already be found that introduce digital twin to wireless networks to improve overall performance. For example, in [9], the authors

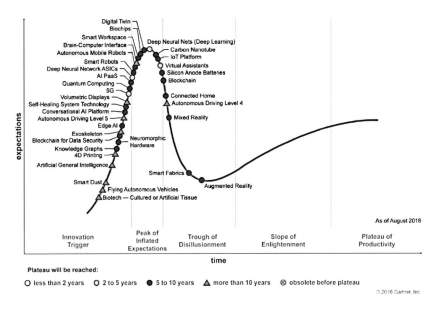

Fig. 1.2 The hype cycle of emerging technologies and digital twin [6]

proposed a new architecture, that is, digital twin edge networks, by integrating digital twins with blockchain to provide secure and accurate edge computing in multi-access edge computing systems.

1.2 Digital Twin Concepts, Features, and Visions

Recent studies have provided a series of definitions for digital twin. In our work, we categorize the definitions into three types: virtual mirror model–based definitions, computerized model–based definitions, and integrated system–based definitions. In the virtual mirror model–based definitions, a digital twin is defined as the virtual representation of a physical product, process, or system [10]. However, in this definition, the interaction between physical objects and digital space is neglected. The status change of physical objects can hardly be reflected by the digital model after its creation. In the computerized model–based definitions [11], digital twins are treated as computerized models, which can be simulation processes or a series of software. The performance of physical objects can be improved through prediction, real-time control, and optimization. In the integrated system–based definitions, digital twins are regarded as an integration of physical objects, virtual twins, and related data [12]. The real-time interaction between physical objects and virtual twins is emphasized in this definition. The virtual twin collects the update information from physical objects and makes corresponding predictions of their future state.

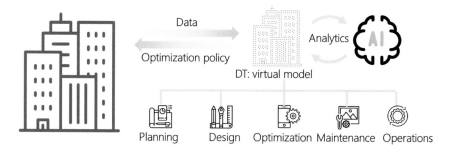

Fig. 1.3 Illustration of the digital twin model

Based on the above analysis, we can give a comprehensive definition of digital twin. Digital twins are accurate digital replicas of real-world objects across multiple granularity levels, and the real-world objects can be devices, machines, robots, industrial processes, or complex physical systems. As shown in Fig. 1.3, the digital twin in virtual space is composed of a virtual model, corresponding running data, and analytical tools. The interaction between real-world objects and digital space is bidirectional: on one side, physical systems transmit real-time data to the virtual space for building digital twin models; on the other, digital twins analyse the collected data, update digital twin models, and provide physical objects with optimization policies to improve their operation performance.

A complete digital twin model consists of three components: the data, the model, and software, as shown in Fig. 1.4.

- *Data, the foundation:* Because the establishment of digital twins relies heavily on historical and real-time running data, data are the foundation of the entire digital twin model. The physical systems contain data, the running principle of the entities, and the controller that can adjust the running of the physical systems.
- *The model, the core component:* The models (both theory models and data-driven models) are the core component of digital twins. The theory models are constructed based on the principles from the physical systems. The data-driven models are trained by collecting the large amounts of historical and real-time running data from the physical systems. A variety of techniques, including artificial intelligence (AI) and data processing, can be used to train the data-driven model. Learning models from data is an iterative process, where the models are trained and updated constantly, as a self-learning process.
- *Software, the essential carrier:* Software is the carrier of digital twins and provides the interface between physical systems and digital space. The digital twin models, composed of algorithms, code, and software, are implemented through developing corresponding software. The functions of representation, diagnosis, prediction, and decision are deployed in a software-defined way that provides the physical controller with optimized command to improve the running performance of the physical systems.

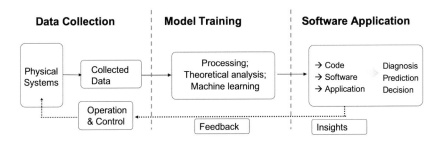

Fig. 1.4 Digital twin: data–model–software

The digital twin provides precise mapping from the physical world to digital space, with real-time interactions. Digital twin mitigates the huge gap between physical space and digital space through continuous synchronization and updates. The features of digital twins can be summarized as follows.

- *Precise mapping*: Digital twin establishes the mapping between physical objects and digital space. The historical data and current running status of physical objects are synchronized to the digital space for further processing and analysis. Based on the transmitted data, digital twins can completely reflect the status of physical objects and establish full mapping between the physical space and digital space. The mapping should completely reflect the full state of physical objects, with low mapping error.
- *Real time*: Different from conventional simulation and modelling technologies, digital twins keep synchronizing with physical objects in real time. The collected data are computed on the digital twin side to extract the corresponding status and build the model of physical objects. Communication is also executed continuously to update the digital twin models. Thus, real-time edge computing should be implemented to ensure the timeliness of digital twins.
- *Distributed*: The physical objects of digital twins can be sensors, IoT devices, and physical systems. In digital twin–assisted wireless networks, multiple physical entities are distributed across the network. The digital twins of these entities are also distributed among different edge servers. In such cases, distributed AI techniques are required to model digital twins from distributed physical objects.
- *Intelligent*: In addition to reflecting the running status of physical objects through real-time data, digital twins also incorporate running models of physical objects. Intelligent techniques, especially AI algorithms, can be used to build digital twin models by processing and analysing the large amounts of running data. With the help of the constructed models, digital twins can provide physical systems with optimization, decisions, and predictions. For example, in intelligent transportation, the digital twin of road traffic can help drivers decide on the optimal

path by analysing real-time traffic conditions and predicting traffic conditions in the near future.

- *Bidirectional*: The interactions between digital twins and physical objects are bidirectional: physical objects transmit and update their running data to digital twins, and digital twins provide physical objects with optimization decisions. The real-time feedback from digital twins to physical objects is one of the unique characteristics of digital twin technology.

Digital twins can provide physical systems with the following benefits through optimization, prediction, and automation processes.

- *Higher performance:* Digital twin can improve system performance through making optimal decisions and executing operations to adjust and control the running of physical systems. In addition, the planning and design of physical systems, such as industrial equipment and healthcare products, can be implemented by using a digital twin to simulate the real-world running performance. Thus, the performance of physical systems can be improved by using digital twins to collect their real-time data and instruct their further operation.
- *Closer monitoring:* Digital twins should copy the complete status of physical objects. To achieve this, physical objects continuously update their running data at a digital twin server. Digital twin servers can closely monitor physical objects in a proactive way that could not be achieved by human operators or conventional monitoring instruments. Digital twins integrate historical data, real-time data, and predicted data to track past states, monitor the current status of physical objects, and predict future conditions for making optimization decisions.
- *Lower maintenance costs:* By collecting the real-time states of physical objects and systems, digital twins can provide optimal maintenance strategies for executing real-time operations. Conventional scheduled maintenance is usually determined in the design phase of physical objects, which leads to high costs and low maintenance efficiency. Digital twins can perform predictive maintenance based on the real-time status of physical objects, which considerably reduces maintenance costs.
- *Increased reliability:* Digital twins can be used to provide virtual tests or simulations of the running of physical objects or systems. The timely digital twin–assisted assessment can improve the quality of physical objects and enhance the reliability of real-world systems.
- *Lower physical system failure risks:* Certain operations can cause physical system failure, resulting in high loss and damage. Digital twins can provide more realistic and accurate simulations for various operations. The simulated environment provided by digital twins is identical to real-world conditions. Thus, the operations of real-world objects and systems can be precisely explored, simulated, and tested to avoid detrimental impacts on physical systems.

With the aforementioned benefits, digital twins can be applied in a variety of scenarios to enhance physical system performance.

- *Smart manufacturing:* Traditional manufacturing faces the problems of limited production efficiency and long product life cycles. Although historical data and simulation have been applied, non–real-time interactions are involved, as well as the low utility of real-time data. The connection between physical objects and virtual space is the key challenge for smart manufacturing in the era of Industry 4.0. Digital twins can integrate physical systems with digital space by analysing huge amounts of historical and real-time data throughout a product life cycle. The results from digital space can provide instructions to the products and processes of physical space. In this way, operation instructions can be optimized, and the quality and efficiency of the manufacturing process can be improved.

- *Aviation:* Aviation was the first area to apply digital twins to practical scenarios. Digital twin has improved data processing and problem diagnosis in aircraft maintenance, risk prediction, construction, and self-maintenance. The full life cycle status of aircraft can be monitored and evaluated by digital twins through real-time data analysis. Real-time operations, such as for flight routes and predictive maintenance, can be determined and optimized with the help of digital twins. However, some challenges remain to be addressed in this area. Because aircraft need precise control and instructions to ensure flight safety, digital twins should be modeled with high accuracy. However, unreliable communications between physical aircraft and digital twin servers can decrease robustness and increase the error rate of digital twin models, which is one of the critical challenges for digital twin–assisted aviation.

- *Intelligent transportation:* Conventional transportation systems have faced serious problems such as traffic jams and accidents. With the assistance of electronic sensors, data analysis, and intelligent control, digital twins can help to improve traffic management and optimize transportation planning efficiency. Traffic accidents can be effectively predicted and avoided through the real-time monitoring and analysis of digital twins. In addition, digital twins can make optimal maintenance decisions for transportation facilities, based on simulations and evaluations of their usage. However, digital twins also face several challenges in this area. The dynamic and time-varying traffic environment poses critical challenges for establishing accurate transportation digital twin models. In addition, the large amounts of data that contain the running states and information of smart vehicles must be transmitted to digital twin servers. Unreliable communication conditions and high transmission latency increase the difficulty of building perfect digital twins.

- *Healthcare:* With the help of IoT devices, it is possible to establish a digital twin for the human body by using IoT sensors and intelligent monitoring equipment to detect a patient's health condition. The physiological status, medication input information, emotional state, and lifestyle of a patient can be collected and analysed in real time by a digital twin. The full range of medical care can be provided by closely monitoring the patient's status and predicting their future health condition. In addition, digital twins can be used in short-term scenarios, such as in remote surgery. For example, experts can obtain real-time feedback by performing operations on the patient's digital twin and identifying potential emergencies that could occur during real-world surgery. Moreover, digital twins can play an important role in the monitoring, management, and maintenance of medical devices. However, given the sensitive nature of patient information, the privacy and security of this data must be treated seriously. Emerging technologies such as privacy computing and blockchain have the potential to improve the protection level of sensitive data.

- *6G Networks:* 6G is predicted to realize the visions of global coverage, enhanced efficiency, fully connected intelligence, and enhanced security. With such visions, tremendous amounts of data must be processed at the edge of the network to provide ultra-low–latency services. In addition, security and privacy issues in data processing need to be addressed. The emergence of digital twins opens up new possibilities to address these challenges for 6G networks. Digital twin technology could be used in various ways to improve the performance of 6G networks. For example, in 6G terahertz communication, digital twin can be used to model, predict, and control signal propagation to maximize the signal-to-noise ratio. In addition, the real-time mirror of physical systems can help to mitigate unreliable and long-distance communications between end users and servers in 6G networks. Digital twins can bridge the huge gap between physical systems and digital space in 6G networks, which can enhance the robustness of wireless connectivity and the intelligence of connected devices. Moreover, network facilities, such as mobile cell towers, can be monitored, planned, and maintained by using digital twins to simulate and evaluate their real-time status.

1.3 Book Organization

This book aims to provide a comprehensive view of digital twin, including fundamentals, visions, and applications. As an emerging technology, digital twin will play an important role in various fields, including manufacturing, transportation, and future networks. However, the high requirements of digital twin, such as ultra-low latency, huge amounts of data transmission, and distributed processing, pose new challenges to the implementation and application of this technology.

To help us confront these challenges, this work provides a comprehensive theory of digital twin and discusses enabling technologies and applications of this paradigm. We first present the fundamental principles of digital twin, including concepts, architectures, features, and visions. Next, we provide digital twin modeling methods and digital twin networks. We also discuss a number of enabling technologies for digital twin, including AI, edge computing, and blockchain. Moreover, we discuss research opportunities for digital twin in the emerging scenarios such as 6G, unmanned aerial vehicles, and Internet of Vehicles. This book is organized as follows. Section 2 presents digital twin models and digital twin networks. Section 3 discusses the use of AI in digital twin, including deep reinforcement learning and federated learning. Section 4 describes the integration of digital twin with edge computing and presents in detail the new architecture of digital twin edge networks. The application of edge intelligence for digital twin is also introduced. Incorporating blockchain into digital twin is discussed in Section 5. Section 6 details the application of digital twin in 6G networks. Finally, Sections 7 and 8 describe the application of digital twin to unmanned aerial vehicles and Internet of Vehicles.

Chapter 2
Digital Twin Models and Networks

Abstract A digital twin (DT) model reflects one or a group of physical objects that exist in a complex real system to a virtual space. By interconnecting and coordinating multiple independent models, a DT network (DTN) can be built to map the associations and interactions between physical objects. In this chapter, we present DT models in terms of modelling frameworks, modelling methods, and modelling challenges. Then we elaborate the concept of DTNs and compare it with the concept of DT. The communication mechanisms, application scenarios, and open research issues of DTNs are then discussed.

2.1 Digital Twin Models

The main goal of DT technology is to reflect the physical world into a virtual space composed of DT models corresponding to different physical objects. As the basic element for realizing the DT function, a DT model describes the characteristics of objects in multiple temporal and spatial dimensions. More specifically, the model always contains the physical object's geometric structure, real-time status, and background information and can further include a fully digital representation of the object's interaction interfaces, software configuration, behavioural trends, and so forth. In this section, we review the DT modelling framework and introduce three categories of DT modelling approaches. Moreover, we discuss the challenges and unexplored problems of DT modelling.

The framework acts as a roadmap of the DT modelling process, which guides the twin system planning, digital model design, mapping step implementation, and performance evaluation. In particular, the DT modelling framework breaks down the complex modelling process into explicit parts and helps to elucidate the factors or interactions that affect the mapping accuracy. Several previous works have focused on the DT modelling framework.

A general and standard framework for DT modelling was first built by Grieves [4]. In the framework, the DT model was described in three dimensions, that is, the

© The Author(s) 2024

Y. Zhang, *Digital Twin*, Simula SpringerBriefs on Computing 16,

https://doi.org/10.1007/978-3-031-51819-5_2

physical entity, the virtual model, and the connection between the physical and virtual parts. This framework has been widely applied to guide DT model construction for industrial production.

Inspired by Grieves' general framework, several studies have extended its hier-archical structure. For instance, Liu *et al.* [13] presented a four-layer DT modelling framework consisting of a data assurance layer, a modelling calculation layer, a DT function layer, and an immersive experience layer. Schroederet *et al.* [14] further introduced a framework composed of a device layer, a user interface layer, a web service layer, a query layer, and a data repository layer. Compared with Grieves' framework, the frameworks proposed in [13] and [14] consider the interactions be-tween the users and DT models, in addition to physical–virtual interactions, and further emphasize the function of DT.

Different from the aforementioned studies, Tao *et al.* [5] described the DT model architecture from the perspective of components. The authors proposed a five-dimensional DT model that encompasses the physical part, the virtual part, the data, connections, and services. This multidimensional framework fuses the data from both the physical and virtual aspects into a DT model to comprehensively and accurately capture the features of the physical objects. Moreover, the framework can encapsulate DT functions, such as environment detection, action judgement, and trend prediction, into the unified management of virtual systems and the on-demand use of twin data. The framework in [5] mainly highlights the influence of system characteristics composed of physical data, virtual data, service data, and historical experience on both virtual twins and mapping services. Due to the completeness of the architecture in terms of DT system composition and element association analysis, it has become one of the main references in the DT modelling process.

2.1.1 DT Modelling Methods

Along with the advancement of wireless technologies and the ever-increasing de-mand for ubiquitous Internet of Things (IoT) services, a vast number of intercon-nected smart devices and powerful infrastructures have spread around the world, making physical systems much more complex and diverse while adding significant difficulty to modelling physical objects in virtual space. In response to this prob-lem, three types of DT modelling approaches catering to different physical systems and application requirements have been introduced: a specific modelling method limited to a given application field, a multidimensional modelling method with mul-tiple functions, and a standard modelling approach for generic DT models. Figure 2.1 compares these modelling approaches in terms of their applicable scenarios, advantages, and disadvantages.

Specific modelling refers to a method that selects only the parameters most rel-evant to a given application scenario as the input data for the mapping and uses a unique mathematical model for the object's model construction. For instance, in [15], specific DT modelling for a power converter was described as a real-time prob-

	Specific model	General model	Multi-dimensional model
Scenarios	Specific scenario	Commonly used in most scenarios	• Multiple sub-models • Satisfy diverse requirements of a scenario
Methods	Mathematical model	CAD, DELMIA, FlexSim, Automod	Machine model + Sensing & control model + Statistical model + Machine learning model
Advantages	• Low computation resource consumption • Satisfy specific requirements	• Apply to large-scale complex systems • High scalability • Interact with each other	• Match various requirements of a scenario
Disadvantages	• Not applicable to large-scale complex systems	• No obvious shortcomings	• Low scalability • Difficult to interact with each other

Fig. 2.1 Comparison between different DT modelling approaches

abilistic simulation process with stochastic variables developed through polynomial chaos expansion. The most important consideration in this scenario was the energy efficiency of the converter, so only parameters relevant to this objective were used as input data. Consequently, this converter DT model has a significantly lower computational cost than similar models. Similarly, in [16], a DT for structural health monitoring based on deep learning was proposed to perform real-time monitoring and active maintenance for bridges. In this work, the modelling method focused on mechanical calculus and quality assessment.

Benefiting from its specificity, the specific DT modelling approach can theoretically be perfectly adapted to given environmental characteristics and to meet particular application requirements. However, due to dynamic and nonlinear relations between physical objects, in most complex application scenarios it is very challenging to generate accurate system mapping in virtual space through a single mathematical model. The use of multidimensional DT modelling based on associated mathematical models seems a promising way to address this challenge.

The multidimensional modelling approach decomposes the entire DT model construction into several submodel building processes, where each submodel corresponds to an explicit task requirement or mapping function. Some work has adopted this modelling approach. In [17], the individual combat quadrotor unmanned aerial vehicle (UAV) model is constructed as a combination of multiple specific models, including a geometric model, an aerodynamics model, a double closed-loop control behaviour model, and a rule model. In the DT modelling process, a submodel can use specific software, extract parts of parameters, and reflect an aspect of the physical objects. For instance, the three-dimensional (3D) modelling software SolidWorks has been leveraged to build the geometric model of the quadrotor UAV. Position coordinates, inertia moment, materials, and other parameters of the UAV are set according to the actual physical conditions. The aerodynamics model is used to realize the flight of the UAV model in the virtual environment. Moreover, a double closed-loop cascade control behaviour model is adopted to ensure the accurate mapping of the UAV. Through iterative optimization, feedback, updates, and adjustment of the UAV's position and altitude parameters, a highly efficient and accurate DT model is ultimately achieved.

In modern industrial manufacturing, 3D DT models of products can be used as experimental objects in production process optimization. Taking into account the diverse attributes of the products, the authors in [18] constructed a 3D printed DT model, using a mechanistic model, a sensing and control model, and a statistical model together with big data and machine learning technology. In the proposed modelling scheme, each model has a specific use. The mechanistic model is used to estimate the metallurgical attributes, such as the transient temperature field, solidification morphology, grain structure, and phases present. The sensing and control model is then used to connect multiple sensors, such as an infrared camera for temperature measurement, an acoustic emission system for capturing surface roughness, and an in-situ synchrotron for monitoring selected geometric features. Besides the models, machine learning technology is leveraged to compare the expected results of the mechanical models with the results obtained from big data sets to determine strategies for tuning the modelling approach.

Although multidimensional modelling can match various application requirements arising in complex environments, the coordination between heterogeneous submodels is not always efficient. Especially for some scenarios with dynamic and variable requirements, this multidimensional but fixed modelling approach can have poor scalability and is not suitable for flexible DT deployment. To address this problem, we can resort to a general modelling mechanism. The general model is always oriented to the multiple requirements of a certain application field. Based on the premise of comprehensively extracting the characteristic parameters of the physical objects, a general but complex DT mapping system is constructed by using standard software tools. For instance, in the field of industrial manufacturing, there are several instances of software development in general modelling for production design and operation analysis, such as Modelica [19], AutoMod [20], FlexSim [21], and DELMIA [22].

Modelica is an open, object-oriented, equation-based general modelling language that can cross different fields and easily model complex physical systems, including mechanical, electronic, electric, hydraulic, thermal, control, and process-oriented subsystems models. Unlike Modelica, AutoMod is a computer modelling software package based on the AutoMod simulation language. It is mainly suitable for establishing DT models of material handling, logistics, and distribution systems. AutoMod contains a series of logistics system modules, such as conveyor modules, automated access systems, and path-based mobile equipment modules. It covers 3D virtual reality animation, interactive modelling, statistical analysis, and other functions.

Compared with the previous two general modelling tools, FlexSim and DELMIA have broader application scenarios. FlexSim is the only simulation software that utilizes a C++ integrated development environment in a graphical model environment. It is designed for engineers, managers, and decision makers to test, evaluate, and visualize proposed solutions on operations, processes, and dynamic systems. It has complete C++ object-oriented function, super 3D virtual reality, and an easy-to-follow user interface. Moreover, due to its excellent flexibility, FlexSim is customized for almost all industry modelling scenarios. Another modelling tool, DELMIA, focuses on a combination of front-end system design data and the resources of a manufacturing site and thus reflects and analyses entire manufacturing and maintenance processes through a 3D graphics simulation engine. The acquired digital data encompasses the visibility, accessibility, maintainability, manufacturability, and optimum performance of the production process. This tool provides a group of production-related libraries and smart visualizers in digital space for factory management.

Scientific studies have also addressed general modelling methods. In [23], Schluse *et al.* proposed a DT modelling technology called Virtual Testbeds that provides comprehensive and interactive digital reflections of operation systems in various application scenarios. Moreover, these testbeds consistently introduce new structures and processes for simulations throughout their life cycle. In [24], Bao *et al.* designed a model-based definition technology to provide digital information carriers and twin images for industrial products during their design, manufacturing, maintenance, repair, and operation phases. As a typical general DT modelling technology, model-based definition technology fuses multidimensional model parameters into a single data source and enables industrial production and services to operate concurrently in virtual space.

2.1.2 DT Modelling Challenges

Although several DT modelling methods for industrial production, modern logistics, and wireless communications have been introduced in both academia and industry, there are still challenges to be addressed to achieve generalization, flexibility, and robustness of the modelling process.

First, there is a lack of standardized frameworks that guide DT modelling in its various forms. A complete DT system is usually composed of a variety of heterogeneous subsystems. These subsystems differ significantly in their functions, structures, and elements. Therefore, different DT models, including geometric models, simulation models, business models, and so forth, need to be used to describe the respective subsystems. Although various modelling frameworks have been developed, none can simultaneously satisfy different virtual modelling requirements while accurately mapping the entire physical system. A standardized modelling framework is expected to be able to cope with various application requirements in different scenarios and stages and realize interoperability among the multiple heterogeneous submodels it contains. However, the design and implementation of this framework remain an unexplored problem.

The second challenge is how to achieve high accuracy in DT modelling. Traditional DT modelling approaches are based on general programming languages, simulation languages, and software to construct the corresponding models. The model can serve only as a reference for the operation process of the physical system and cannot provide the core data required for virtual model construction with high-precision object descriptions and state prediction. In addition, traditional DT modelling can suffer from poor flexibility, complex configurations, and error proneness.

Finally, how the DT models respond and react in real time to events occurring in the physical space is a critical challenge. In the real world, the characteristics of physical objects, such as their geometric shape, energy consumption, topological relations, and so on, change dynamically. To cope with these changes, the DT modelling should be updated accordingly. However, limited by sensing capability and data transmission capacity, it can be difficult to obtain comprehensive and real-time system state data in practical scenarios. How to perform high-fidelity model updates based on incomplete information acquisition in DT space is a problem worthy of future investigation.

2.2 DT Networks (DTNs)

A DTN is defined as a many-to-many mapping network constructed by multiple one-to-one DTs. In other words, a DTN uses advanced communication technologies to realize real-time information interactions between a physical object and its virtual twin, the virtual twin and other virtual twins, as well as the physical object and other physical objects. A DTN realizes the dynamic interactions and synchronized evolution of multiple physical objects and virtual twins by using accurate DT modelling, communications, computing, and physical data processing technologies. In a DTN, physical objects and virtual twins can communicate, collaborate, share information, complete tasks with each other, and form an information-sharing network by connecting multiple DT nodes. In this section, we first analyse the difference between DT and a DTN. Next, the communications in DTNs are discussed. Further, we depict some typical DTN application scenarios such as manufacturing, sixth-generation

(6G) networks, and intelligent transportation systems. Finally, we point out open research issues related to DTN.

2.2.1 DTN Concepts

Figure 2.2 compares the concepts of a DT and a DTN in terms of application scenarios, composition structure, and mapping relationships.

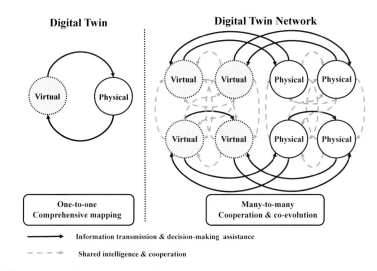

Fig. 2.2 Comparison between a DT and a DTN

First, from the perspective of application scenarios, the concepts of DT and a DTN are different. DT is suitable for reflecting a single independent object, whereas a DTN models a group of objects with complex internal interactions. For example, modelling a building in virtual space through the DT approach helps optimize the entire life cycle of the building in terms of design, maintenance, and so on. The building model depends only on the analysis and decision making according to the building's state data. In contrast, when building a virtual model of an industrial automation production line, a DTN should be used to model and reflect the collaborative relationships between the multiple industrial components involved in the production process.

Second, from the perspective of the operation mode, DT focuses on modelling an individual physical object in virtual space, and a DT model always gathers and processes the object's state information in an independent mode without interacting with other models. Constrained by an individual DT model's information collection and processing capabilities, the constructed object model might not be accurate enough, while both the time and energy consumption of this construction process can be high. In contrast to DT, a DTN collaborates between multiple DTs to model a group of objects. The information of the physical object, the processing capability of the DT model, and some intermediate processing results can be shared among the collaborative DTs. This cooperation approach significantly reduces processing time delays and energy consumption and greatly improves modelling efficiency.

Finally, from the perspective of physical and virtual mapping relationships, DT provides comprehensive physical and functional descriptions of components, products, or systems. The main goal of DT is to create high-fidelity virtual models to reproduce the geometric shapes, physical properties, behaviours, and rules of the physical world. Enabled by DT, virtual models and physical objects can maintain similar appearances as twin brothers and the same behaviour pattern as mirror images. In addition, the model in digital space can guide the operation of the physical system and adjust physical processes through feedback. With the help of two-way dynamic mapping, both the physical object and the virtual model evolve together. Considering the mirroring effect of each physical and logical entity pair, we classify the mapping relationship between physical and virtual space in a DT system as one to one. We then characterize the mapping relationship of a DTN as many to many.

In summary, DT is an intelligent and constantly evolving system that emphasizes a high-fidelity virtual model of a physical object. The mapping relationship between physical and virtual spaces in the DT system is one to one, with high scalability. A DTN is extended as a group of multiple DTs. By applying communications between DTs, a one-to-one mapping relationship can be easily expanded to a DTN. The mapping relationship is also more conducive to network management. Combined with advanced data processing, computing, and communications technologies, DTNs can easily facilitate information sharing and achieve more accurate state sensing, real-time analysis, efficient decision making, and precise execution on physical objects. Compared with DT, a DTN, which uses a network form to build complex large-scale systems, is more reliable and efficient.

2.2.2 DTN Communications

The establishment of a DTN relies on the information exchange and data communication between the physical objects in the real world and the logical entities in virtual space. According to different combinations of communication object pairs, these communications can be divided into three types: physical-to-virtual, physical-to-physical, and virtual-to-virtual communications.

Physical-to-virtual communications can be considered the process of transferring information from a physical system to virtual entities. This type of communication meets the requirements of the DT modelling process for the characteristic parameters of physical objects, and it can also feed back the modelling results to the physical space to guide parameter collection and transmission adjustment. Physical-to-virtual technology mainly uses wide area network wireless communication paradigms, such as LoRa and fifth-generation/6G cellular communications. In these paradigms, the physical objects are wireless terminals connected to a wireless access network through a wireless communication base station that further relays data to a virtual twin connected to the Internet. The communication infrastructures are robust to support real-time interactions between the physical and virtual.

Physical-to-physical communications ensure information interactions and data sharing between physical objects. Various wireless or wired devices, such as sensors, radio frequency identification, actuators, controllers, and other tags, can connect with IoT gateways, WiFi access points, and base stations supporting physical-to-physical communications. In addition, the network connections are enabled by diverse communication protocols, such as wireless personal area networks and Zigbee, and low-power wide area network technologies, including LoRa and Narrowband IoT.

Virtual-to-virtual communications, which logically encompass the virtual space, mirror the communication behaviour in the real physical world. For instance, in the Internet of Vehicles use case, virtual-to-virtual communications refer to data transmission between the DT model entities of the vehicles. Unlike communications between physical vehicles that consume vehicular wireless spectrum resources and radio power, this virtual mode depends mainly on DT servers' computing capability to model data transmission behaviours. Another key benefit of virtual-to-virtual communications is the data transmission modelling, which breaks through the time constraints of the physical world. We note that communications between actual vehicles consume a certain amount of time. However, in virtual space, the same communication behaviour can be completed much more quickly. Thus, we can reflect or simulate a long period of communication behaviour with a low time cost. Furthermore, a given communication behaviour can logically occur in virtual space earlier than it actually occurs in physical space. The effect of logical communications can be leveraged to guide resource scheduling in the real world. Edge intelligence, which consists of artificial intelligence–empowered edge computing servers, is a critical enabling technology for achieving virtual-to-virtual communications. Edge servers thus provide the necessary computing capability for channels' model construction and data transmission while artificial intelligence learns the characteristics of the physical network and adjusts the communication modelling strategies.

2.2.3 DTN Applications

With the development of DT technology, many application scenarios using DTN to assist process management and policy adjustment have emerged, such as smart manufacturing, 6G networks, and intelligent transportation systems.

Subject to the high costs of updating production, traditional manufacturing has problems with low production efficiency and outdated product designs. The introduction of DTNs in new smart manufacturing can effectively address these problems. For the factory production line, by establishing a virtual model of the entire line, the production process can be simulated in advance and problems in the process found, thereby achieving more efficient production line management and process optimization. Moreover, in the real production process, the virtual twin of the factory can be continuously updated and optimized, including the DT model of factory construction, product production, industrial equipment life prediction, system maintenance, and so forth. DTNs that match production requirements are helpful for achieving efficient digital management and low-cost manufacturing.

6G networks aim to integrate a variety of wireless access mechanisms to achieve ultra-large capacity and ultra-small distance communications. In reaching these goals, 6G networks could face challenges in terms of security, flexibility, and spectrum and energy efficiency. The emergence of DTNs provides opportunities to overcome these challenges. DTNs enable 6G networks to realize innovative services, such as augmented reality, virtual reality, and autonomous driving. A DTN can virtually map a 6G network. The virtually reflected 6G network collects the traffic information of the real communication network, implements data analysis to discover data traffic patterns, and detects abnormal occurrences in advance. The 6G network uses the information fed back from the virtualized network to prepare network security protection capabilities in advance. In addition, by collecting and analysing the communication data in the DTN, communication patterns can be determined. Then, by reserving communication resources, the demand and supply of data delivery services can be automatically achieved.

In recent years, the urban transportation system has experienced road network congestion and frequent traffic accidents. DTNs leverage multidimensional information sensors, remote data transmission, and intelligent control technology to provide information assistance services for intelligent traffic management and autonomous vehicle driving. First, a DTN provides a virtual vision of the transportation system, helping to dispatch traffic and optimize public transportation services. Next, by processing massive amounts of real-time traffic information, the virtual system of a DTN can accurately predict traffic accidents and thus help avert them.

2.2.4 Open DTN Research Issues

As an emerging technological paradigm, DTNs have demonstrated strong physical system mapping and information assistance capabilities. Both DTN operation technologies and application scenarios have been studied, but further research questions remain.

Security is one of the key research issues of DTNs. A DTN is a complex system composed of virtual mappings of various networks and objects. This complex structure makes its security difficult to protect. Moreover, information sharing within virtual networks can raise security concerns. In a DTN, a pair of twins has a bidirectional feedback relationship. Even if the physical system in the real world is well secured, an attacker can easily change the parameters of the virtual model or the data fed back by the virtual model. Such attacks are particularly harmful to data-sensitive applications such as intelligent transportation systems and medical applications.

DTNs rely on real-world information, whose gathering process can cause privacy leaks. For example, in intelligent medical care, the virtual modelling of the human body needs to collect various types of biological information and monitor the patient's daily activities. In treatment, sensitive data can be sent to and processed on edge servers. Edge service operators can share these data with other companies without user consent, which increases the risk of privacy breaches. How to balance data utilization and privacy protection turns out to be a critical challenge for DTN exploration.

Another research issue to consider is resource scheduling. The construction of a DTN consumes a variety of heterogeneous resources, including sensing resources for information collection, communication resources for data transmission, computing resources for modelling processing, and cache resources for model preservation. These resources jointly affect the efficiency and accuracy of DTN operation. The way to optimize resource scheduling is worthy of future investigation.

Chapter 3
Artificial Intelligence for Digital Twin

Abstract Artificial intelligence (AI) is a promising technology that enables machines to learn from experience, adjust to environments, and perform humanlike tasks. Incorporating AI with digital twin (DT) makes DT modelling flexible and accurate, while improving the learning efficiency of AI agents. In this chapter, we present the framework of AI-empowered DT and discuss some key issues in the joint application of these two technologies. Then, we introduce the incorporation paradigms of three AI learning approaches with DT networks.

3.1 Artificial Intelligence in Digital Twin

AI is a branch of computer science that enables learning agents to perform tasks that typically rely on human intelligence. Nowadays, the blooming of AI technology has brought powerful capabilities in environmental cognition, knowledge learning, action decision, and state prediction to smart machines, vehicles, and various types of Internet of Things (IoT) devices.

However, despite great advancements led by AI for industry, transportation, healthcare, and other areas, AI is not always glamorous. In fact, the AI learning process consists of continuous interactions between agents and the environmental system. The agents make decisions and take actions according to the current observed environment states, and these actions then react to and change the environment states, which triggers a new round of agent learning until the process finally converges. The interactive learning approach, which relies on real physical systems, is often costly and inefficient. For instance, when applying AI directly to real vehicles to train autonomous driving policies, vehicles can cause traffic accidents. Another example is leveraging AI to optimize the operation of cellular networks. Due to the large scale of cellular networks and their many subscribers, it takes a long time for AI agents to obtain feedback on state changes after performing actions, which seriously undermines AI learning effectiveness. Incorporating AI with simulation software seems a feasible approach to speed up the system feedback for AI actions. However,

Y. Zhang, *Digital Twin*, Simula SpringerBriefs on Computing 16,
https://doi.org/10.1007/978-3-031-51819-5_3

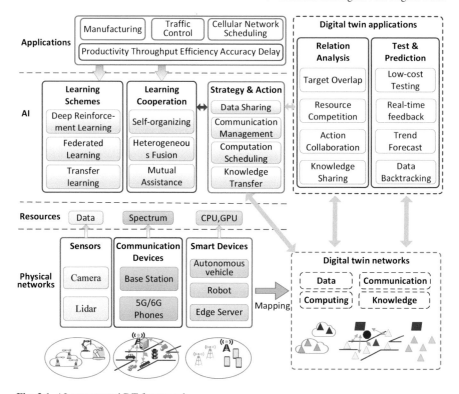

Fig. 3.1 AI-empowered DT framework

due to nonlinear factors and uncertainty, it is hard to build a high-fidelity simulation environment for a highly dynamic and complex system. Thus, the strategies and actions learned in a simulation environment cannot be directly deployed to machines in the real world.

To cope with this problem, we resort to DT technology. DT mirrors the forms, states, and characteristics of physical objects in the real world with high fidelity and real time into virtual space. This mirror model eases our cognition of complex physical systems and makes operations on virtual entities equivalent to those on physical ones. Moreover, by leveraging the precise reflection capability of DT and the intelligent adaptability of AI, the combination of DT and AI can benefit both parties. On the one hand, with the aid of DT, AI learning methods can obtain high-fidelity state information from physical objects for model training, verify the effect of the learning strategy at low cost, and implement the life cycle management of complex systems. On the other hand, AI learning can continuously monitor the accuracy of DT models, dynamically adjust the DT mapping mechanism, and maintain the consistency between virtual space and physical space.

To fully explore the benefits of incorporating AI and DT, we present the framework of AI-empowered DT shown in Fig. 3.1. This framework is mainly composed of two types of networks, namely, physical networks and DT networks. The physical networks are composed of various types of physical devices and different types of resources served or consumed by these devices. As ubiquitous devices in the physical networks, sensors such as cameras and lidars collect the real-time state from the physical environment. The state data carry the characteristics of the real world, such as the operating conditions of industrial equipment and driving behaviour of smart vehicles, which are useful for failure detection and traffic planning. Another type of device involves communication infrastructures and user terminals, for instance, cellular radio base stations and mobile phones. The data interaction between these devices mostly adopts wireless communications, which consume spectrum resources. Thus, managing the communication equipment mainly involves scheduling of channel resources. Furthermore, in the physical network, smart devices play an important role in providing computation resources. Smart devices such as autonomous vehicles, edge servers, and robots can be equipped with very powerful CPU and GPU computing capabilities, compared to handheld user devices. For such data-intensive and computationally intensive tasks, performing local computations on user equipment can consume excessive energy and bring about long delays. Catering to this problem, these tasks can be offloaded to edge service–enabled smart devices for efficient processing.

The data, communication, and computing resources mentioned above can be scheduled to serve various types of tasks in the physical network. However, the highly dynamic topology of mobile devices and communication interference arising in the physical environment pose significant challenges to resource efficiency and application performance. More specifically, the mobility and dynamic topology of devices make environmental data collection more difficult. In addition, due to wireless interference, the received data will deviate from the sender's original data, which can lead to erroneous environmental cognition and resource scheduling decisions. To address these challenges, we turn to DT technology and formulate DT networks.

A DT network is a mapping of a physical network in a virtual space that consists of virtual twins of all the physical units on the physical side. Data, spectrum, and computing resources contained in the DT network form logical entities that can be freely decomposed and flexibly combined. In addition, the resources in the DT network include some of the knowledge and experience that have been already gained and cached, such as the channel history states and known bandwidth allocation strategy in previous radio resource management. Since the DT network operates in a virtual space, there is no interference or error in the information interaction between DT entities, and the coordination of heterogeneous resources can also break the constraints of node locations and realize resource supply and demand services between distant nodes. In addition, based on historical information and knowledge, future network status trends can be accurately predicted, thereby facilitating effective resource management.

Based on the DT networks formulated, two promising types of applications can be achieved. The first is a variety of relational analysis in complex systems and highly dynamic environments, including objective overlap testing in distributed optimization, competition for limited resources by multiple business nodes, action cooperation among multiple nodes, and knowledge sharing collaboration among a group of machine learning agents. The other type of DT application is strategy testing and future state prediction. DT can provide low-cost policy verification in virtual space and obtain real-time result feedback. Moreover, during the forecasting process, the time axis can be easily and flexibly adjusted, allowing for efficient trend forecasting and data retrospectives.

The AI module for scheduling resources can be divided into two parts. The first part involves learning schemes to determine the architecture as well as the components of AI models, and it can be mainly classified into deep reinforcement learning (DRL), federated learning (FL), and transfer learning (TL). Among these schemes, DRL is of an architecture combining the neural networks of deep learning (DL) and the decision model of reinforcement learning (RL). Based on DRL, FL is a multi-agent DRL framework that can protect the privacy of each agent. TL is a novel concept that aims to utilize the original model to construct a new model to speed up convergence.

The second part of the AI module involves learning cooperation relationships that indicate the cooperation types between learning agents, including self-organizing, heterogeneous fusion, and mutual assistance. These relationships can be further classified into individual learning and cooperative learning. Individual learning always converges faster than cooperative learning, since it does not experience a time delay in information interaction. However, a lack of global information about a system can cause the convergence point to be suboptimal. In contrast, collaborative learning can usually achieve more accurate decision performance, but it often requires longer convergence times, especially for large-scale complex systems.

3.2 DRL-Empowered DT

3.2.1 Introduction to DRL

In earlier years, machine learning methods represented by DL and RL were widely used to solve various problems in networks. DL aims to construct deep neural networks to identify characteristics from the environment, while RL aims to take optimal actions to obtain maximal rewards. More specifically, DL enables machines to imitate human activities such as hearing and thinking and to solve complex pattern recognition problems, making great progress in AI-related technologies. RL allows agents to imitate the capacity of humans making decisions based on the current environment. However, both DL and RL have their drawbacks. For example, DL cannot explain decisions it has already made, and RL cannot identify high-dimensional states of the environment well. Combining DL with RL to design a new machine learning framework called DRL is a promising approach to address the above problem. DRL combines the perceptive ability of DL with the decision-making ability of RL. Moreover, DRL can learn control strategies directly from high-dimensional raw data, which is much closer to human learning compared to previously designed AI approaches.

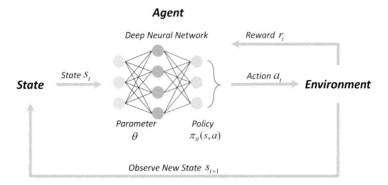

Fig. 3.2 DRL framework

Essentially, DRL is applied to sequential decision making, which can be mathematically formulated as a Markov decision process (MDP). The DRL framework is shown in Fig. 3.2. In each time slot t, the agent observes the current environment state s_t and uses its policy to select an action a_t. A policy can be considered a mapping from any state to an action. After the action a_t is performed, the environment moves to state s_{t+1} in the next time slot with transition probability $P(s_{t+1}|s_t, a_t)$. In addition, a corresponding reward $r_t = R(s_t, a_t)$ is obtained via the immediate reward function, which is the evaluative feedback of the action taken. Given a stationary and Markovian policy π, the next state of the environment, s_{t+1}, is completely determined by the current state, s_t. In this context, the current policy together with the transition probability function determines the long-term cumulative reward. Assuming $\tau = (s_t, a_t, s_{t+1}, a_{t+1}, \cdots, s_T, a_T)$ is a trajectory from an MDP, the long-term cumulative reward can be defined as

$$G(\tau) = \sum_{i=0}^{T-t} \gamma^i R(s_{t+i}, a_{t+i}), \tag{3.1}$$

where $\gamma \in (0, 1]$ is the discount factor that measures the importance of the future reward and T is the length of an episode. For a continuous MDP, we have $T \longrightarrow \infty$. In an MDP, the key issue is to find the optimal policy that maximizes the long-term cumulative reward.

3.2.2 Incorporation of DT and DRL

As a promising AI technology, DRL provides a feasible method for solving complex problems in unknown environments. However, there are still challenges to be resolved in the process of DRL learning and implementation, which are discussed below.

High cost of the trial-and-error learning process: As a zero-knowledge experimental learning method, DRL maximizes the cumulative discounted reward by learning optimal state–action mapping policies through trial and error. However, in some application scenarios, especially in traffic safety–sensitive Internet of Vehicles applications and smart medical care related to patients' lives, the cost of trial and error is too high to be acceptable.

Frequent data transmission in learning: A large amount of state data needs to be input into the DRL system to train models and draw action strategies. For example, the channel spectrum status and real-time communication requirements of users are input for radio resource scheduling. Rapid and dynamic changes in environmental status and user requirements result in intensive data transmission and frequent state updates. Furthermore, as the dimensionality of the input data increases, so too does the time taken for the learning process to reach the convergence. Thus, we find that it is difficult for the DRL method to meet the needs of delay-sensitive business scenarios such as the driving action control of autonomous vehicles and communication management in interactive multimedia applications.

Interaction barriers between multiple agents in distributed DRL: Distributed DRL uses multiple agents to obtain the optimal action policy based on the environmental status. These agents can accelerate the learning process by sharing information when collaboratively working towards a common learning target. However, when the agents use wireless communication to share learning information, wireless signal fading and spectrum interference can lead to transmission errors and retransmission, which not only cause extra communication costs, but can also undermine training efficiency and learning convergence.

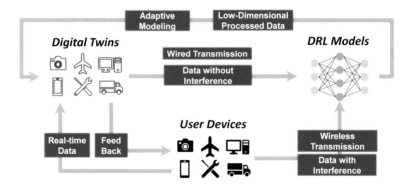

Fig. 3.3 Cooperation of DT and DRL

To address the above challenges, we turn to DT technology. Figure 3.3 illustrates how DT and DRL can cooperate to improve learning efficiency. First, since DT creates a high-fidelity virtual map of physical objects, DRL algorithms applied in the real world can be trained in the DT space. Different from the real training process in physical space, the trial-and-error process in DT training does not have unacceptable consequences, such as damage or injury to objects or humans due to wrong decisions. Second, the agents of DRL can obtain physical system states from the DT models without relying on communications between the agents and the physical objects, reducing data transmission delays. Compared with traditional DRL implemented in the physical space, the DRL model on the DT side can be trained for more rounds per unit time and converges faster. Finally, by modelling the DT of DRL agents on DT servers, the actual information interaction between agents in the physical space can be mapped to the information sharing between DT servers or within one server in virtual space. This virtual-to-virtual agent communication enables reliable information sharing between two agents and does not consume physical communication resources.

On the DRL side, we note that the features, functions, and behaviours of physical objects are often high dimensional, making it difficult to describe them directly in the DT modelling process. With the help of DRL, these high-dimensional data are extracted and refined by neural networks into lower-dimensional data that are easier to process. Furthermore, DRL can help handle some of the unique problems of DT, such as DT placement and DT migration algorithms, and make DT technology adaptable to different time-varying environments.

Numerous recent studies have investigated the cooperation of DRL and DT. Among these works, the resource management of sixth-generation (6G) networks has attracted much attention from researchers. In [25], the authors considered the dynamic topology of the edge network and proposed a DT migration scenario. They adopted a multi-agent DRL approach to find the optimal DT migration policy by considering both the latency of updating DT and the energy consumption of data transmission. In [26], the authors proposed an intelligent task offloading scheme assisted by DT. The mobile edge services, mobile users, and channel state information were mapped into DT to obtain real-time information on the physical objects and radio communication environments. Then, a reliable mobile edge server with the best communication link quality was selected to offload the task by training the data stored in the DT with the double deep-Q learning algorithm. In [27], the authors proposed a mobile offloading scheme in a DT edge network. The DT of the edge server maps the state of the edge server, and the DT of the entire mobile edge computing system provides training data for offloading decisions. The Lyapunov optimization method was leveraged to simplify the long-term migration cost constraint in a multi-objective dynamic optimization problem, which was then solved by actor–critic DRL. This solution effectively diminishes the average offloading latency, the offloading failure rate, and the service migration rate while saving system costs with DT assistance.

DT technology and DRL can be seamlessly fused to achieve intelligent manufacturing. In [28], the authors proposed a DT- and RL-based production control method. This method replaces the existing dispatching rule in the type and instance phases of a micro smart factory. In this method, the RL policy network is learned and evaluated by coordination between DT and RL. The DT provides virtual event logs that include states, actions, and rewards to support learning. In [29], the authors proposed the automation of factory scheduling by using DT to map manufacturing cells, simulate system behaviour, predict process failures, and adaptively control operating variables. Moreover, based on one of the cases, the authors presented the training results of the deep Q-learning algorithm and discussed the development prospects of incorporating DRL-based AI into the industrial control process. By applying the DRL method, process knowledge can be obtained efficiently, manufacturing tasks can be arranged, and optimal actions can be determined, with strong control robustness.

In addition to the above work, previous studies have applied DT and DRL to emerging applications. In [30], the authors analysed a multi-user offloading system where the quality of service is reflected through the response time of the services; they adopted a DRL approach to obtain the optimal offloading decision to address the problem of edge computing devices overloading under excessive service requests owing to the computational intensity of the DT-empowered Internet of Vehicles. In [31], the authors discussed the feedback of traditional flocking motion methods for unmanned aerial vehicles (UAVs) and proposed a DT-enabled DRL training framework to solve the problem of the sim-to-real problem restricting the application of DRL to the flocking motion scenario.

3.2.3 Open Research Issues

Although the cooperation of DRL and DT has shown great potential in some scenarios, there are still problems that warrant investigation. The first problem is resource scheduling. The volume of data of physical objects in DT is huge, and the deployment of DRL at the edge also requires computing resource services. Therefore, reducing redundant data and designing lightweight DRL models are significant issues in the combination of DT and DRL.

Another issue is environmental dynamics. The DT modelling process can involve a dynamic and time-varying environment, with a wide variety of physical objects, and the data and computing requirements required for the corresponding modelling processing can also differ. In addition, the high-speed movement of physical objects and the dynamic changes of wireless channels will further exacerbate the uncertainty of environmental characteristics. Although DRL can provide an optimal strategy for DT resource scheduling, a continuously and dynamically changing environment can seriously undermine learning efficiency. Therefore, improving the flexibility and adaptability of DRL to dynamic DT modelling is an important issue to be addressed.

3.3 Federated Learning (FL) for DT

3.3.1 Introduction to FL

The proliferation of AI learning techniques has provided unprecedented powerful applications to areas including smart manufacturing, autonomous driving, and intelligent healthcare. With these diverse AI applications, two critical challenges have emerged that must be addressed. The first challenge is learning scalability. In a system with many widely distributed nodes, using a traditional centralized AI mechanism in the learning process can generate significant amounts of data to be collected and in overhead transmission, creating a great burden on the processing capability of a few centralized agents. Another challenge centres around privacy protection. The system states or data resources gathered for learning related to factory production techniques, route navigation preferences, and an individual's personal physical condition invariably contain sensitive information, requiring a strong privacy guarantee.

FL has been widely regarded as an appealing approach to address the above challenges. FL is a privacy-protected model-training technology with an emphasis on leveraging distributed agents to collect data and leverage local training resources. Unlike centralized AI, which depends purely on the capability of a few central agents, in FL multiple geodistributed agents perform model training in parallel without sharing sensitive raw data, thus helping ensure privacy and reducing communication costs.

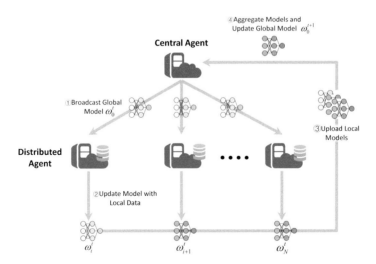

Fig. 3.4 Main flow of the FL process

Figure 3.4 shows the main flow of the FL process. First, a central agent initializes a global model, denoted as ω_0, and broadcasts this model to the other distributed agents. Then, after each distributed agent receives ω_0, it takes locally collected data to update the parameters of this model and achieves a local model that minimizes the loss function, defined as

$$F(\omega_i^t) = \sum_{x_i \in D_i} f(\omega_i^t, x_i) \Bigg/ |D_i|, \qquad (3.2)$$

where ω_i^t is the local model of agent i in learning iteration t, and D_i is the local data set of agent i. This loss function is used to measure the accuracy of the local model and guide the model update in a gradient descent approach, which is written as

$$\omega_i^{t+1} = \omega_i^t - \xi \cdot \nabla F(\omega_i^t), \qquad (3.3)$$

where ξ is the learning step. Next, each distributed agent uploads its local model to the central agent and waits for an aggregation step, which can be written as

$$\omega_0^{t+1} = \sum_{j=1}^{N} \alpha_i \cdot \omega_i^t \Bigg/ N, \qquad (3.4)$$

where α_i is the coefficient of agent i and N is the number of collaborating learning agents. When the aggregation is completed, the central agent will republish the updated global model to the distributed agents. The iterations repeat in this manner until the global model converges or reaches a predetermined accuracy.

3.3.2 Incorporation of DT and FL

Although FL is a promising paradigm that enables collaborative training and mitigates privacy risks, its learning operation still has several challenges and limitations.

Complexity and uncertainty of model characteristics: Large-scale dynamic systems usually have diverse features that correlate with each other, which means it is very difficult for FL to extract them from system events. Moreover, during the learning operation, unplanned events such as weather changes, traffic accidents, and equipment failures, can further confuse the training inputs and undermine model convergence.

Asynchrony between heterogeneous cooperative agents: As a distributed AI framework, FL leverages multiple geographically distributed agents to train their local models in parallel and then aggregates a parametric model in a central agent. There is heterogeneity in the training environment where each agent is located in terms of the number of physical entities, the size of the region, the frequency of event changes, and the differences in agents' processing capacity. This heterogeneity makes it hard to synchronize the aggregation of FL across multiple distributed agents. Although previous works have been devoted to the design of asynchronous FL mechanisms, most of them have improved the learning convergence at the cost of model accuracy. How to achieve both learning efficiency and model precision is still an open question.

Interaction bottleneck between collaborative agents: Considering the distributed training and central aggregation characteristics of FL, frequent interactions are required between the client agents and the central agent, especially for learning systems with high-dimensional feature parameters and highly dynamic environments. In such a case, where wireless communications are used to realize the interactions between agents, the efficiency of local model aggregation and global model distribution can be severely undermined due to the data transfer bottleneck caused by the limited wireless spectrum and disturbed

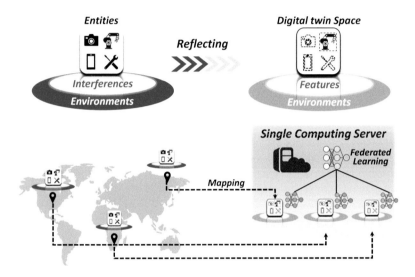

Fig. 3.5 Benefits of applying DT in FL

To address the above challenges, we turn to DT technology. Figure 3.5 illustrates the benefits of applying DT in FL. First, reflecting complex physical entities and environments into DT space can eliminate unnecessary interference factors, thereby helping FL to mine the core features of the system and further explore their interrelationships. Second, for the problem of asynchronous heterogeneous training regions, using a mirrored virtual environment built by DT to replace all or part of the regional systems affected by slow response can greatly improve these regions' local model convergence speeds. The training between regions is thus synchronized, and both learning efficiency and accuracy can be achieved. Finally, the DT mappings of multiple regions can be constructed on a single computing server, and the real data communications between the agents located in different regions in the physical space can be mapped to the interactions between multiple learning processes in the virtual space. Therefore, DT can free the collaborative agents in FL from the constraints of physical communication resources.

We note that the many benefits provided by DT to FL depend on the ability of the twin models in the virtual space being able to map physical entities and networks accurately and in real time. Due to the potential dynamics of physical networks, the DT mapping strategy needs to be adjusted accordingly. Considering the large-scale and distributed characteristics of the physical entities, using FL to optimize the mapping strategy seems an appealing approach. More specifically, in the integration of DT and FL, DT mapping accuracy can be included as an element of the learning reward, and the parameters of the DT mapping strategy can be added to the learning action space.

Recently, research attempts have focused on applying DT with FL. Among these works, the Industrial IoT (IIoT), which enables manufacturers to operate with massive numbers of assets and gain insights into production processes, has turned out to be an important application scenario. In [32], the authors intended to improve the quality of services of the IIoT and incorporated DT into edge networks to form a DT edge network. In this network, FL was leveraged to construct IIoT twin models, which improves IIoT communication efficiency and reduces its transmission energy cost. In [33], the authors used DT to capture the features of IIoT devices to assist FL and presented a clustering-based asynchronous FL scheme that adapts to the IIoT heterogeneity and benefits learning accuracy and convergence. In [34], the authors focused on resource-constrained IIoT networks, where the energy consumption of FL and digital mapping become the bottleneck in network performance. To address this bottleneck, the authors introduced a joint training method selection and resource allocation algorithm that minimizes the energy cost under the constraint of the learning convergence rate.

In preparation for the coming 6G era, DT technology and FL can be seamlessly fused to trigger advanced network scheduling strategies. In [9], the authors presented an FL-empowered DT 6G network that migrates real-time data processing to the edge plane. To further balance the learning accuracy and time cost of the proposed network, the authors formulated an optimization problem for edge association by jointly considering DT association, the training data batch size, and bandwidth allocation. In [35], the authors applied dynamic DT and FL to air–ground cooperative 6G networks, where a UAV acts as the learning aggregator and the ground clients train the learning model according to the network features captured by DTs.

In the area of cybersecurity, blockchain has emerged as a promising paradigm to prevent the tampering of data. Since both the ledger storage of blockchain and the model training process of FL are distributed, blockchain can be introduced into DT-enabled FL. In [36], the authors utilized blockchain to design a DT edge network that facilitates flexible and secure DT construction. In this network, a double auction–based FL and local model verification scheme was proposed that improves the network's social utility. In [37], the authors proposed a blockchain-enabled FL scheme to protect communication security and data privacy in digital edge networks, and they introduced an asynchronous learning aggregation strategy to manage network resources.

In addition to the above work, previous studies have applied DT and FL to emerging applications. In [38], the authors used the COVID-19 pandemic as a new use case of these two technologies and proposed a DT–FL collaboratively empowered training framework that helps the temporal context capture historical infection data and COVID-19 response plan management. In [39], the authors applied these two technologies to edge computing–empowered distribution grids. A DT-assisted resource scheduling algorithm was proposed in an FL-enabled DT framework that outperforms benchmark schemes in terms of the cumulative iteration delay and energy consumption.

3.3.3 Open Research Issues

The incorporation of FL with DT is a promising way to improve learning efficiency while guaranteeing user privacy. However, there are still unexplored questions in the joint application of these two technologies. The first question worth investigating is the operation matching between DT and FL. The training process of FL requires many iterations, which consume massive computing resources and generate a certain time delay. Since DT modelling also depends on intensive computation, competition for resources arises between DT and FL. Effective resource scheduling is thus a critical research challenge. Moreover, the key advantage of DT is the ability to accurately map the physical world into virtual space in real time. When using FL to improve DT modelling accuracy, how to make the slow iterative learning direct the DT mapping strategy in a timely manner is still a problem for future research.

Another unexplored question concerns privacy. To reflect physical systems and objects fully and accurately, DT modelling inevitably needs to extract massive amounts of system data and user information, which can lead to privacy leakage. On the other hand, the use of FL is an attempt to protect users' private information. How to ensure privacy protection while improving the accuracy of DT modelling is also a challenge to be addressed.

3.4 Transfer Learning (TL) for DT

3.4.1 Introduction to TL

In traditional distributed intelligence networks, multiple machine learning agents equipped on edge servers, smart vehicles, and even powerful IoT devices, work independently. In some application scenarios, multiple agents in similar environments can learn with the same goal. If these agents start training at different times, agents that start later may learn their strategies from scratch. A complete training process always incurs a great deal of resource consumption and long training delays, posing a critical challenge for resource-constrained devices serving delay-sensitive computing tasks.

TL, which is a branch of AI with low learning costs and high learning efficiency, provides a promising approach to meet these challenges. Unlike the traditional machine learning agent that tries to learn a new mission from scratch, a TL agent receives prior knowledge from other agents that have performed similar or related missions, and then starts learning with the aid of this knowledge, thus achieving faster convergence and better solutions.

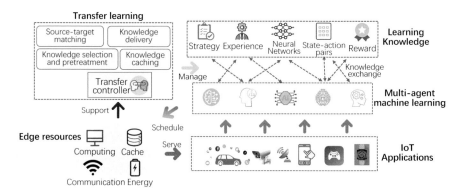

Fig. 3.6 TL framework

Figure 3.6 illustrates the TL framework. At the bottom of this figure are shown various types of modelling training and strategy learning tasks generated by IoT devices. Multiple agents with TL capabilities are deployed to handle these tasks. We note that FL-inspired learning is a gradual process that consists of continuous environment awareness, constant action exploration, and persistent strategy improvement. As the learning proceeds, valuable knowledge, such as neural network parameters, state–action pairs, action exploration experience, and the evaluation of existing strategies, is generated and recorded. This knowledge not only is the basis for the learning of the local agent in subsequent stages, but also can be shared with other agents, which can then jump directly from the initial learning stage, without any experience, to an intermediate stage with certain prior knowledge.

In the FL framework, a transfer controller module manages the sharing process, including the pairing of the transfer source and target agents, knowledge building and pretreatment, the knowledge data delivery, and the caching among the agents. It is worth noting that edge resources play a vital role in the FL framework. On the one hand, these resources can serve in IoT applications, such as vehicular communications, popular video caching, and sensing image recognition, while multi-agent machine learning is leveraged for resource scheduling. On the other hand, we resort to TL to improve machine learning efficiency and reduce scheduling time costs. However, the knowledge sharing process can create the need for extra communication, computing, and cache resources. Thus, there exists a trade-off in resource allocation, that is, whether to use the resources to directly enhance IoT application performance or for learning efficiency improvement and service delay reduction.

TL can offer many benefits in multi-agent distributed learning scenarios, the main advantage being the reduction in training time of the target agent of the knowledge sharing process. The shared prior knowledge can effectively guide the agent to quickly converge to and reach optimal action strategies without time-consuming random exploration. In addition, TL can save training resource consumption. Each training step requires analysis and calculation. A faster training process means fewer steps, as well as lower computing and energy resource consumption. Moreover, for machine learning approaches that record large amounts of state–action pairs, the reduced training process provided by the TL also reduces the record sizes, thereby saving on cache resources.

3.4.2 Incorporation of DT and TL

Despite the benefits provided by TL, unaddressed challenges remain in TL scheme implementation, especially in application scenarios with multiple associated heterogeneous agents. Due to the associations between such agents, multiple TL node pairs can be formed. Thus, the first challenge is the choice of transferring source when the target mission has multiple potential knowledge providers. For example, when multiple UAVs are agents in training terrain models based on sensing data, these UAVs hover and cruise at different altitudes and can have overlapping or even the same modelling area. The beneficial prior knowledge of a UAV agent performing a learning mission can exist in multiple neighbouring UAVs. Source determination is a prerequisite before the learning implementation. However, it is difficult to determine the appropriate transferring pairs solely according to the physical characteristics and superficial associations in the physical world. Another challenge is what knowledge should be transferred. The prior knowledge learned by heterogeneous agents can take various forms and provide diverse learning gains between different transferring pairs. Knowledge selection and organization are the basis of effective TL. However, since knowledge is an abstract concept, it is hard to measure and schedule it accurately in physical space.

Incorporating DT with TL is a feasible approach to address the above challenges. In terms of the effect of DT on TL, by leveraging the comprehensive mapping ability of DT from a physical system to virtual space, multi-agents' environmental characteristics, neural network structure, and learning power, as well as their current training stages can be clearly presented in a logical form. This logical representation allows the TL scheduler to find optimal TL source–destination agent pairs based on the similarity of environmental features or the matching of knowledge supply and demand. Moreover, DT models existing in the virtual space are suitable for describing the knowledge attributes acquired by each agent. For example, knowledge can be logically represented as a tuple DT model composed of an owner, information items, the application scope, transfer gains, transfer costs, and other elements.

From the perspective of the role played by TL in the DT process, especially in scenarios of distributed multi-DT models, TL can share the construction experience of the completed DT model, such as the model structure, constituent elements, and update cycle, with the DT models that have been or have yet to be formed. This knowledge transfer scheme greatly shortens DT construction delays and improves DT model accuracy. Moreover, since DT processes consume considerable communication and computing resources, TL can also be used in several similar DT environments to reuse resource scheduling strategies.

TL has been used in many areas to improve the efficiency of distributed learning. For instance, in [40], the authors proposed a deep uncertainty–aware TL framework for COVID-19 detection that addresses the problem of the lack of medical images in neural network training. In [41], the authors introduced a TL-empowered aerial edge network that uses multi-agent machine learning to draw optimal service strategies while leveraging TL to share and reuse knowledge between UAVs to save on resource costs and reduce training latency. In [42], TL was used in action unit intensity estimation, where known facial features were inherited in new estimation scenarios at minimal extra computational cost.

Along with the development of DT technology, a few studies have been dedicated to the incorporation of DT and TL. In [43], the authors focused on anomaly detection in dynamically changing network functions virtualization environments. They used DT to measure a virtual instance of a physical network in capturing real-time anomaly–fault dependency relationships while leveraging TL to utilize the learned knowledge of the dependency relationships in historical periods. In [44], the authors introduced a DT and deep FL jointly enabled fault diagnosis scheme that diagnoses faults in both the development and maintenance phases. In this scheme, the previously trained diagnosis model can be migrated from virtual space to physical space for real-time monitoring. Considering that DT models are usually customized for specific scenarios and could lack sufficient environmental adaptability, the authors in [45] leveraged TL to explore an adaptive evolution mechanism that improves remodelling efficiency under the premise of limited environmental information.

3.4.3 Open Research Issues

As recent emerging technologies, DT and TL, as well as their incorporation, still have open research issues to be explored. The first issue concerns knowledge transfer between heterogeneous training models. Training models can differ among TL agents, in terms of their learning methods, neural network structures, and knowledge cache organization. Although DT can describe these training models logically and consistently in virtual space, during TL implementation, how to preprocess and match the knowledge between source and target agents to improve the transfer effect is still a key challenge.

The second issue involves resource scheduling in DT-empowered TL. Various types of resources play a key role in TL for knowledge data delivery, storage, and processing, and DT's model building and updating also consume these resources. Competition for constrained resources can thus take place during cooperation between FL and DT. How to coordinate resource scheduling between the two and improve the efficiency of knowledge transfer while ensuring modelling accuracy is therefore also a key question to be addressed.

Finally, an issue to be considered is DT construction that adapts to TL operations. TL usually occurs between multiple agents distributed in a large-scale system, whereas DT systems always construct models on a small number of centralized servers. How to solve the contradiction between the distributed architecture of TL and the centralized construction of DT requires further exploration.

Chapter 4
Edge Computing for Digital Twin

Abstract Mobile edge computing is a promising solution for analysing and processing a portion of data using the computing, storage, and network resources distributed on the paths between data sources and a cloud computing centre. Mobile edge computing thus provides high efficiency, low latency, and privacy protection to sustain digital twin. In this chapter, we first introduce a hierarchical architecture of digital twin edge networks that consists of a virtual plane and a user/physical plane. We then introduce the key communication and computation technologies in the digital twin edge networks and present two typical cooperative computation modes. Moreover, we present the role of artificial intelligence (AI) for digital twin edge networks, and discuss the unique edge association problem.

4.1 Digital Twin Edge Networks

4.1.1 Digital Twin Edge Network Architecture

In traditional cloud computing–assisted digital twin modelling, the centralized server collects data and constructs twin mappings of the physical components, which leads to large communication loads. In this context, digital twin edge networks, a new paradigm that integrates mobile edge computing (MEC) and digital twin to build digital twin models at the network edge, has emerged as a crucial area. In digital twin edge networks, the edge nodes—for example, base stations (BSs) and access points—can collect running states of physical components and develop their behaviour model along with the dynamic environment. Furthermore, the edge nodes continuously interact with the physical components by monitoring their states, to maintain consistency with their twin mappings. Hence, the networking schemes (i.e. decision making, prediction, scheduling, etc.) can be directly designed and optimized in the constructed digital twin edge networks, which improves the efficiency of networking schemes and reduces costs. To better understand the internal logic

© The Author(s) 2024
Y. Zhang, *Digital Twin*, Simula SpringerBriefs on Computing 16,
https://doi.org/10.1007/978-3-031-51819-5_4

of digital twin edge networks, we first present a hierarchical architecture of these networks that consists of a virtual plane and a user/physical plane, as shown in Fig. 4.1.

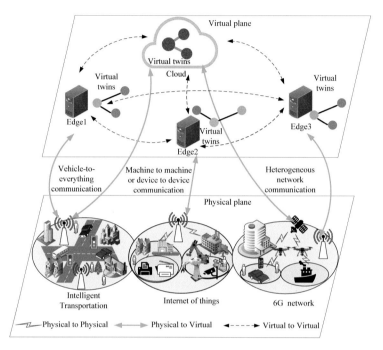

Fig. 4.1 A hierarchical architecture of digital twin edge networks

The user/physical plane is distinguished by typical digital twin application scenarios, such as an intelligent transportation system, the Industrial Internet of Things (IIoT), and sixth-generation (6G) networks. The virtual plane generates and maintains the virtual twins of physical objects by utilizing digital twin technology at the edge and on cloud servers. Specifically, devices in the user/physical plane include vehicles, sensors, smart terminals, and so forth. These devices need to synchronize their data with the corresponding virtual twins in real time through wireless communication technologies. Meanwhile, these devices also accept feedback from their virtual twins for instantaneous control and calibration. Therefore, mobile edge networks are expected to provide communications and computations that satisfy the main requirements of low latency, high reliability, high speed, and privacy and security preservation, to support real-time interactions between physical and virtual planes.

4.1.1.1 Communications

Communications between physical and virtual planes in typical digital twin edge network scenarios can be summarized as follows.

- *Intelligent transportation systems:* In recent years, urban transportation systems have faced such problems as traffic jams and traffic accidents. Digital twin edge networks can provide a virtual vision of the transportation system that can help to manage traffic and optimize public transportation service planning efficiency [46]. For example, traffic accidents can be effectively predicted and avoided by processing the massive amounts of real-time transportation information in the virtual plane [47]. Digital twin edge networks can also offer new opportunities for maintaining transportation facilities. By simulating the usage of transportation facilities in the virtual plane, facility malfunctions can be predicted in advance, which helps managers to schedule appropriate maintenance actions.

  Vehicle-to-everything communications allow vehicles to communicate with other vehicles and their virtual twins via wireless links, which can be realized by dedicated short-range communications and fifth-generation/6G communications [48]. In digital twin edge network–enabled intelligent transportation systems, vehicles' running states and perceived environmental information need to be transmitted to the virtual plane to update the virtual twins. However, it is challenging to guarantee strict data transmission delays, since vehicles move at high speeds. A detailed communications design must be carefully considered for physical plane and virtual plane interactions in such a dynamic network environment.

- *Internet of Things (IoT):* With the increasing scale of the IoT, digital twin is one of the most promising technologies enabling physical components be connected with their virtual twins in digital space by using different sensing, communication, computing, and software analytics technologies, to provide configuration, monitoring, diagnostics, and prognostics for maintaining physical systems [49, 50]. For example, in manufacturing, digital twin edge networks can be utilized for different aspects of manufacturing to improve production efficiency and reduce product life cycles [51, 52]. When designing parts, their full life cycle can be simulated through a virtual model, and design defects can be found in advance to realize accurate parts design. In factory production lines, through a virtual model of the entire production line, the production process can be simulated in advance and problems in the process found, to achieve more efficient production line management and process optimization. Additionally, in the health domain, digital twin edge networks can be utilized to establish twin patients. The twin patients can collect patients' physiological status and life style, medication input data, and data about the patients' emotional changes over time. Thus, twin patients can enable medical experts to provide patients with a full range of medical care and even accurately predict changes.

  Machine-to-machine and device-to-device (D2D) communications are enabling technologies for the digital twin edge network–empowered IoT [53]. Physical components can form clusters and transmit shared status data to the corresponding virtual twins by reusing the unoccupied uplink spectrum resources. Machine-to-machine and D2D communications can improve data transmission rates during physical plane and virtual plane interactions. However, privacy and security protection of the information in virtual twin formation is a critical issue, since some core data, such as users' personal information, must be continuously updated for the virtual twins, and malicious attackers could intercept this information through wireless communications. Hence, privacy and security protection mechanisms need to be designed for physical and virtual plane interactions in the IoT.

- *6G networks:* 6G networks aim to realize ultra-high-capacity and ultra-short-distance communications, go beyond best effort and high-precision communications, and converge multiple types of communications [27]. Thus, 6G networks can face challenges in security, spectral efficiency, intelligence, energy efficiency, and affordability. The emergence of digital twin edge networks introduces opportunities to overcome these challenges. Digital twin edge networks provide corresponding virtual twins of 6G network components, which can collect traffic information on the entire network and use data analysis methods to discover network traffic patterns and detect abnormal traffic in advance. 6G networks use the information fed back from the virtual twins to make preparations in advance to improve network performance. In addition, by collecting and analysing the communication data in networks, rules of communication can be discovered to automate demand and provide services on demand. Since communication demand can be predicted in advance, the information can be fed back to the 6G networks to reserve resources, such as spectrum resources.

  The interactions between the physical and virtual planes in digital twin–empowered 6G networks demand high data rates. Small cell communication is an efficient solution for improving spectral efficiency by deploying heterogeneous infrastructures, such as pico and micro BSs [54]. In small cell communication, all BSs are equipped with rich computational resources and are responsible for generating and maintaining the virtual twins of physical objects in the cells. Additionally, intelligent communication infrastructures, such as reconfigurable intelligent surfaces [55] and unmanned aerial vehicles [56, 57], can be leveraged to realize interactions between the physical and virtual planes.

4.1.1.2 Computations for Resource-Intensive Tasks in the Virtual Plane

Beyond communications with low latency and high reliability, the resource-intensive tasks executed by digital twin edge networks require large amounts of computational resources. The virtual plane in the hierarchical architecture of digital twin edge networks consists of multiple distributed edge servers and central cloud servers. Specifically, central cloud servers have strong processing, caching, and computing capabilities. Resource-intensive tasks that focus on computation speed and centralized processing can be deployed on central cloud servers. Through cloud servers, large amounts of data can be processed in a short time (a few seconds), to provide powerful digital twin services for physical objects. In addition, the cloud architecture facilitates the organization and management of large numbers of connected physical objects and virtual twins, as well as the combination and integration of real-time data and historical experience.

In addition, edge servers have computing (i.e. CPU cycles) and caching resources distributed on the paths between data sources and the cloud computing centre that can analyse and process a portion of the data from both physical objects and virtual twins. Edge servers can be deployed in the network infrastructure, such as at BSs, roadside units, wireless access points, gateways, and routers, or they can be mobile phones, vehicles, and other devices with the necessary processing power and computing and storage capabilities. Considering the proximity of edge servers to physical objects, delay-sensitive tasks can be deployed on edge servers to provide digital twin services for users with high efficiency.

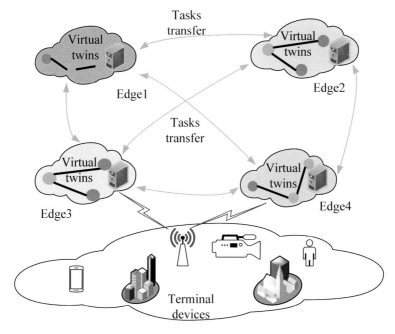

Fig. 4.2 Cooperative edge computing in digital twin edge networks

4.1.2 Computation Offloading in Digital Twin Edge Networks

In digital twin edge network scenarios, data processing and analysis require great amounts of computing resources. Nevertheless, most criteria cannot be met by edge computing, due to the limited capacity of edge servers. For example, when an edge node has many computing tasks with a long task queue, it can easily create high latency. Cooperative computation is an approach for offloading computing tasks to other nodes that have free computing resources, to reduce task processing latency. According to different cooperation methods among the nodes, the following two cooperative computation modes can be used.

4.1.2.1 Cooperative Edge Computing

In cooperative edge computing, as shown in Fig. 4.2, if other edge nodes have free computing resources, they should share in the computing tasks of the overloaded edge nodes. It is very important for multiple edge nodes to maintain workload balance and provide low-latency computing services, particularly when a digital twin edge network provides services for time-sensitive scenarios, as in intelligence transportation systems.

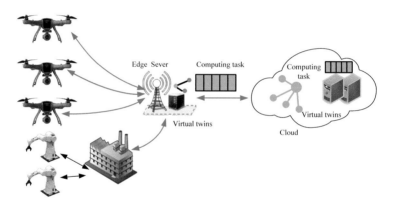

Fig. 4.3 Cooperative cloud-edge-end computing in digital twin edge networks

4.1.2.2 Cooperative Cloud-Edge-End Computing

As shown in Fig. 4.3, cooperative cloud-edge-end computing is necessary to meet the demand for large-scale computations and AI for real-time modelling and simulation in digital twin edge networks. Edge servers process the data that need to be responded to in real time. The cloud server provides strong computing power and the integration of various types of information. The interaction between edge nodes and the cloud in real time can solve the problem of data heterogeneity for the cloud. Cooperative cloud-edge-end computing can provide low-latency computation, communications, and virtual twin continuous updating for digital twin edge networks. In addition, when the storage resources of the edge nodes are insufficient, the cloud can store part of the data and transmit them to the client through the network when needed, which saves storage resources on the edge servers.

In this section, we first present a hierarchical architecture of digital twin edge networks that consists of a virtual plane and a user/physical plane. Then, we illustrate key communications and computation technologies between the physical and virtual planes in the typical digital twin edge network scenarios and present two cooperative computation modes.

4.2 AI for Digital Twin Edge Networks

The integration of digital twin with AI [58] opens up new possibilities for efficient data processing in applying digital twins in 6G networks. MEC, one of the key enabling technologies for 6G, can considerably reduce system latency by executing computations based on AI algorithms at the edge of the network. AI-empowered MEC has been widely investigated for accomplishing edge intelligence tasks such as computation offloading, content caching, and data sharing. In [59], the authors proposed an AI-empowered MEC scheme in the IIoT framework. In [60], the authors proposed an intelligent content caching scheme based on deep reinforcement learning (DRL) for an edge computing framework. AI can significantly improve the construction efficiency and optimize the running performance of digital twin edge networks. The system model, communication model, and computation model of AI-empowered digital twin edge networks are as follows.

4.2.1 System Model

4.2.1.1 AI-Empowered Network Model

We consider the AI-empowered digital twin edge network shown in Fig. 4.4. Our wireless digital twin network system comprises three layers: a radio access layer (i.e. end layer), a digital twin layer (i.e. edge layer), and a cloud layer. The radio access layer consists of entities such as mobile devices and vehicles that have limited computing and storage resources. Through wireless communications, these entities connect to BSs and request services provided by network operators. In the digital twin layer, some BSs are equipped with MEC servers to execute computation tasks, while other BSs provide wireless communication services to end users. The digital twins of the physical entities are modelled and maintained by the MEC servers. Since the number of entities in the physical layer is much larger than the number of MEC servers in the digital twin layer, an MEC server can maintain multiple digital twins of physical entities. In the cloud layer, cloud servers are equipped with large amounts of computing and storage resources. Tasks that are computation sensitive or require global analysis can be executed in the cloud layer.

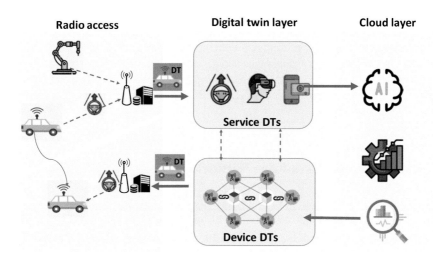

Fig. 4.4 The architecture of wireless digital twin networks

Since digital twins reproduce the running of physical entities, maintaining the digital twins of massive devices consumes a large number of resources, including computing resources, communication resources, and storage resources. To relieve the resource limitation in the edge layer, we model digital twins as one of two types: a device digital twin or a service digital twin. The device digital twin is a full replica of the physical devices, which includes the information of the hardware configuration, the historical running data, and real-time states. The device digital twin for user u_i can be expressed as

$$DT^f(u_i) = \Theta(\mathcal{D}_i, S_i(t), \mathcal{M}_i, \Delta S_i(t+1)), \tag{4.1}$$

where $\mathcal{D}_i$ is the historical data of user device i, such as the configuration data and historical running data. The term $S_i(t)$ represents the running state of device i, which consists of r_1 dimensions and varies with time, and it can be denoted as $S(t) = \{s_i^1(t), s_i^2(t), ..., s_i^{r_1}(t)\}$. The term $\mathcal{M}_i$ is the behaviour model set of u_i, which consists of r_2 behaviour dimensions, and $\mathcal{M}_i = \{m_i^1, m_i^2, ..., m_i^{r_2}\}$, and $\Delta S_i(t+1)$ is the state update of $S_i(t)$ in time slot $t+1$. Taking a meteorological IoT device as an example, $S(t)$ can be the temperature, humidity, wind speed, location, and so on. The behaviour models $\mathcal{M}_i$ can consist of the variation models of the temperature, humidity, and wind speed. In this paper, we mainly focus on the scenarios of device digital twin to conduct our study.

Different from a device digital twin, a service digital twin is a lightweight digital replica constructed by extracting the running states of several devices for a specific application. Similar to (4.1), the service digital twin can be expressed as

$$DT(u_i, \zeta) = \Theta(\mathcal{D}_i(\zeta), S_i^\zeta(t), \mathcal{M}_i^\zeta, \Delta S_i^\zeta(t+1)), \tag{4.2}$$

where ζ is the target service, and $\mathcal{D}_i(\zeta)$, $S_i^\zeta(t)$, M_i^ζ, and $\Delta S_i^\zeta(t+1)$ are the corresponding terms related to the target service ζ. For example, vehicles driving in the same region can be modelled into a specific service digital twin for supporting autonomous driving on a particular stretch of road. In such a case, the service digital twin for autonomous driving collects only the driving information of these vehicles and analyses their driving behaviour to guide them. Depending on the required scale, service digital twins can be constructed on the edge server or the cloud server.

4.2.2 Communication and Computation Model

The communication between end users and edge servers contains the uplink communication for transmitting data from user devices to edge servers and the downlink communication for sending the results from edge servers back to user devices. Note that the size of the results returning to users is much smaller than that of the updated data, so we consider only uplink communication latency in our communication model. The maximum achievable uplink data rate r_{ij} between user i and BS j is given as

$$r_{ij} = W log(1 + \frac{p_{ij} h_{ij}}{W N_0}), \tag{4.3}$$

where h_{ij} denotes the channel power gain of user i, p_{ij} denotes the corresponding transmission power for user i, N_0 is the noise power spectral density, and W is the channel bandwidth. The transmission latency for uploading D_i from user i to BS j can be expressed as

$$T_{ij}^{com} = \frac{D_i}{r_{ij}}. \tag{4.4}$$

The wired transmission latency between BSs is highly correlated to the transmission distance. Let ϕ be the latency required for transmitting one unit of data in each unit distance. Then the wired transmission latency can be written as

$$T_{j_1 j_2}^{com} = \phi \cdot D_j \cdot d(j_1, j_2), \tag{4.5}$$

where D_j is the size of the transmitted data and $d(j_1, j_2)$ is the distance between BSs j_1 and j_2.

We denote the total computation resource of edge server j as F_j. The computation resource of edge server j can be allocated to multiple user devices to maintain their digital twins on server j. Let f_{ij} denote the computation resource assigned to the digital twin of user i. Then the time to execute tasks from user i can be expressed as

$$T_{ij}^{cmp} = \frac{D_i}{f_{ij}}, \tag{4.6}$$

where D_i is the size of computation task from user i, $\sum_{i=1}^{N} x_{ij} f_{ij} \leq F_j$, and $x_{ij} = 1$ if $f_{ij} > 0$. Otherwise, $x_{ij} = 0$.

4.2.3 Latency Model

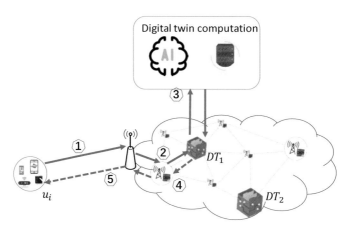

Fig. 4.5 The digital twin construction process

The latency of maintaining a digital twin mainly consists of two parts: the construction delay and the synchronization delay. Figure 4.5 shows the complete process for constructing a digital twin of user u_i. In the beginning, the running data D_i of u_i are transmitted to their nearby BS through wireless communication. Then the nearby BS transmits through wired communication the running data D_i to the digital twin server DT_1 that is responsible for constructing and maintaining the digital twin of u_i. The digital twin server DT_i runs the computation to process and analyse the received data and builds a digital twin model for user u_i, as expressed by Eq. (4.1). During the digital twin computation process, AI-related algorithms are used to extract the data features and to train the digital twin model. Finally, the results of the digital twin model are transmitted back to user u_i through wired and wireless communications. The feedback results provide u_i with insights for improving its service quality or running efficiency for specific applications. The system latency consists of the following items.

1. *Wireless data transmission:* In the construction phase of $DT(u_i)$, the historical running data of user i must be transmitted to its digital twin server through its nearby BS. Let D_i denote the size of the historical data to be transmitted. The wireless communication latency T_{ij}^{com} from user i to its BS j can then be calculated according to Eq. (4.4).
2. *Wired data transmission:* The wired transmission time from the nearby BS of u_i to its digital twin server k is

$$T_{jk}^{com} = \phi \cdot D_i \cdot d(j, k). \tag{4.7}$$

The total communication time for transmitting the historical data of u_i to its digital twin server is thus

$$T_{ik}^{com} = T_{ij}^{com} + T_{jk}^{com}. \tag{4.8}$$

3. *Digital twin data computation:* The computation time at digital twin server k is

$$T_{ik}^{cmp} = \frac{D_i}{f_{ij}}. \tag{4.9}$$

The total latency for constructing the digital twin of user i is

$$T_{ik}^{ini} = T_{ik}^{com} + T_{jk}^{com} + T_{ik}^{cmp}. \tag{4.10}$$

The digital twin of user i, that is, $DT(u_i)$, is constructed on its digital twin server DT_k. Then, $DT(u_i)$ must constantly interact with u_i to remain consistent with the running states of u_i. We denote the size of the updated data as ΔD_i. The latency for one update can then be expressed as

$$T_{ik}^{upd} = \frac{\Delta D_i}{r_{ij}} + \phi \cdot \Delta D_i \cdot d(j, k) + \frac{\Delta D_i}{f_{ij}}. \tag{4.11}$$

The synchronization latency in one unit time slot can be written as

$$T_{ik}^{syn} = \frac{1}{\Delta t} T_{ik}^{upd}, \tag{4.12}$$

where Δt denotes the time gap between every two updates.

4.3 Edge Association for Digital Twin Edge Networks

4.3.1 System Model

Due to the dynamic computing and communication resources available through edge servers, the association of digital twins to corresponding servers is a fundamental problem in digital twin edge networks that needs to be comprehensively explored. Moreover, since the federated learning in digital twin edge networks requires multiple communications for data exchange, the limited communication resources need to be optimally allocated to improve the efficiency of digital twins in the associated edge servers. Thus, in this section, we design a digital twin wireless network (DTWN) model and define the edge association problem for digital twin networks. A permissioned blockchain-empowered federated learning framework for edge association is also proposed.

We consider a blockchain- and federated learning–empowered digital twin network model as depicted in Fig. 4.6. The system consists of N end users, such as

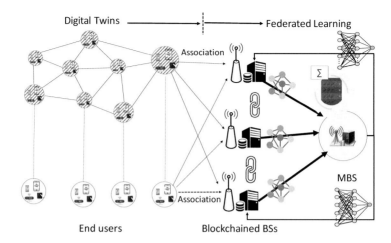

Fig. 4.6 The proposed digital twin wireless network

IoT devices and mobile devices, M BSs, and a macro BS (MBS). The BSs and the MBS are equipped with MEC servers. The end devices generate running data and synchronize their data with the corresponding digital twins that run on the BSs. We use $\mathcal{D}_i = \{(x_{i1}, y_{i1}), ..., (x_{iD_i}, y_{iD_i})\}$ to denote the data of end user i, where D_i is the data size, x_i is the data collected by end users, and y_i is the label of x_i. The digital twin of end user i in the BSs are denoted as DT_i, which is composed of the behaviour model $\mathcal{M}_i$, static running data $\mathcal{D}_i$, and the real-time dynamic state s_t, so that $DT_i = (\mathcal{M}_i, \mathcal{D}_i, s_t)$, where $\mathcal{D}_i$ and s_t are the essential data required to run the digital twin applications. Instead of synchronizing all the raw data to the digital twins, which incurs a huge communication load and the risk of data leakage, we use federated learning to learn model $\mathcal{M}$ from the user data. In various application scenarios, the end users can communicate with other end users to exchange running information and share data, through, for example, D2D communications. Thus, the digital twins also form a network based on the connections of end users. Based on the constructed DTWN, we can obtain the running states of the physical devices and make further decisions to optimize and drive the running of the devices by directly analysing the digital twins.

In our proposed digital twin network model, we use federated learning to execute the training and learning process collaboratively for edge intelligence. Moreover, since the end users lack mutual trust and the digital twins consist of private data, we use permissioned blockchain to enhance the system security and data privacy. The permissioned blockchain records the data from digital twins and manages the participating users through permission control. The blockchain is maintained by the BSs, which are also the clients of the federated learning model. The MBS runs as the server for the federated learning model. In each iteration of federated learning, the MBS distributes the machine learning model parameters to the BSs for training.

The BSs train the model based on the data from the digital twins and returns the model parameters to the MBS.

We use orthogonal frequency division multiple access for wireless transmission in our system. To upload trained local models, all the BSs share C subchannels to transmit their parameters. The achievable uplink data rate from BS i to the MBS is

$$R_i^U = \sum_{c=1}^{C} \tau_{i,c} W^U log_2(1 + \frac{P_{i,c}^U h_{i,c}^U r_{i,m}^{-\alpha}}{\sum_{j \in N'} P_{j,c}^U h_{j,c}^U r_{i,m}^{-\alpha} + N_0}), \quad (4.13)$$

where C is the total number of subchannels, $\tau_{i,c}$ is the time fraction allocated to BS i on subchannel c, and W^U is the bandwidth of each subchannel, which is a constant value. The transmission power is $P_{i,c}^U$ and the uplink channel gain on subchannel c is $h_{i,c}^U$; $r_{i,m}^{-\alpha}$ is the path loss fading of the channel between BS i and the MBS; $r_{i,m}$ is the distance between BS i and the MBS; α is the path loss exponent; N_0 is the noise power; and $\sum_{j \in N'} P_{j,c}^U h_{j,c}^U r_{i,m}^{-\alpha}$ is the interference caused by other BSs using the same subchannel. In the download phase, the MBS broadcasts the global model with the rate

$$R_i^D = \sum_{c=1}^{C} W^D log_2(1 + \frac{P_{i,c}^D h_{i,c}^D}{\sum_{j \in N''} P_{j,c}^D h_{j,c}^D r_{i,m}^{-\alpha} + N_0}), \quad (4.14)$$

where $P_{i,c}^D$ is the downlink power of BS i, $h_{i,c}^D$ is the channel gain between BS i and the MBS, and $\sum_{j \in N''} P_{j,c}^D h_{j,c}^D r_{i,m}^{-\alpha}$ is the downlink inference.

4.3.2 Edge Association: Definition and Problem Formulation

The end devices or users are mapped to the digital twins in the BSs in the DTWN. The maintenance of digital twins consumes a large amount of computing and communication resources for synchronizing real-time data and building corresponding models. However, the computation and communication resources in wireless networks are very limited and should be optimally used to improve resource utility. Thus, the association of various IoT devices with different BSs according to their computation capabilities and states of the communication channel is a key problem in DTWNs. As depicted in Fig. 4.6, the digital twins of IoT devices are constructed and maintained by their associated BSs. The training data and the computation tasks for training are distributed to various BSs based on the association between the digital twins and the

Definition (Edge Association) Consider a DTWN with N IoT users and M BSs. For any user u_i, $i \in N$, the goal of edge association is to choose the target BS $j \in M$ to construct the digital twin DT_i of user i. The association $\langle DT_i, BS_j \rangle$ is denoted as $\Phi(i, j)$. If DT_i is associated with BS j, then $\Phi(i, j) = D_i$, where D_i is the size of the data used to construct DT_i. Otherwise, $\Phi(i, j) = 0$. □

A BS can be associated with multiple digital twins, whereas a digital twin can only be associated with at most one BS; that is, $\sum_{j=1}^{M} \Phi(i, j) = D_i$. We perform edge association according to the datasets D_i of IoT users, the computation capability of the BSs f_j, and the transmission rate $R_{i,j}$ between u_i and BS_j, denoted as

$$\Phi(i, j) = f(D_i, f_j, R_{i,j}). \tag{4.15}$$

The objective of the edge association problem is to improve the utility of resources and the efficiency of running digital twins in the DTWN.

We use the weight matrix $A = [a_{ik}]$ to represent the association relations between the user devices and the digital twin servers, where $a_{ik} = 1$ if the digital twin of user i is maintained by digital twin server k. Otherwise, $a_{ik} = 0$. For example, in Fig. 4.5, since the digital twin of u_i is maintained by DT_1, we have $a_{i1} = 1$ and $a_{i2} = 0$. The association matrix takes the form

$$\begin{bmatrix} a_{11} & a_{12} & a_{1k} & \dots & a_{1M} \\ a_{21} & a_{22} & a_{2k} & \dots & a_{2M} \\ a_{21} & a_{22} & a_{2k} & \dots & a_{2M} \\ . & . & . & \dots & . \\ . & . & . & \dots & . \\ . & . & . & \dots & . \\ a_{N1} & a_{N2} & a_{Nk} & \dots & a_{NM} \end{bmatrix}.$$

Now we start to derive the formulation of the edge association problem. We consider that the gradient $\nabla f(w)$ of $f(w)$ is L-Lipschitz smooth; that is,

$$||\nabla f(w_{t+1}) - \nabla f(w_t)|| \leq L||w_{t+1} - w_t||, \tag{4.16}$$

where L is a positive constant and $||w_{t+1} - w_t||$ is the norm of $w_{t+1} - w_t$. We consider that the loss function $f(w)$ is strongly convex; that is,

$$f(w_{t+1}) \geq f(w_t) + \langle \nabla f(w_t), w_{t+1} - w_t \rangle + \frac{1}{2}||w_{t+1} - w_t||^2. \tag{4.17}$$

Many loss functions for federated learning can satisfy the above assumptions, for example, logic loss functions. If (4.16) and (4.17) are satisfied, the upper bound of the global iterations can be obtained as

$$\mathcal{T}(\theta_L, \theta_G) = \frac{O(log(1/\theta_L))}{1 - \theta_G}, \tag{4.18}$$

where θ_L is the local accuracy $\frac{||\nabla f(w_{t+1})||}{||\nabla f(w_t)||} \leq \theta_L$, θ_G is the global accuracy, and $0 \leq \theta_L, \theta_G \leq 1$. As in [94], we consider θ_L a fixed value, so that the upper bound $\mathcal{T}(\theta_L, \theta_G)$ can be simplified to $\mathcal{T}(\theta_G) = \frac{1}{1-\theta_G}$. If we denote the time of one local training iteration by T_{cmp}, then the computation time in one global iteration is $log(1/\theta)T_{cmp}$, and the upper bound of total learning time is $\mathcal{T}(\theta_G)T_{glob}$.

The time cost in our proposed scheme mainly consists of the following.

1. *Local training on digital twins:* The time cost for the local training of BS i is determined by the computing capability and the data size of its digital twins. The time cost is

$$T_i^{cmp} = \frac{\sum_{j=1}^{K_i} b_j D_{DT_j}}{f_i^C} f^C, \qquad (4.19)$$

where f^C is the number of CPU cycles required to train one sample of data, f_i^C is the CPU frequency of BS i, and b_j is the training batch size of digital twin DT_j.

2. *Model aggregation on the BSs:* The BSs aggregate their local models from various digital twins. The computing time for local aggregation is

$$T_i^{la} = \frac{\sum_{j=1}^{K_i} |w_j|}{f_i^C} f_b^C, \qquad (4.20)$$

where $|w_j|$ is the size of the local models and f_b^C is the number of CPU cycles required to aggregate one unit of data. Since all the clients share the same global model, $|w_1| = |w_2| = \dots = |w_j| = |w_g|$. Thus the time cost for local aggregation is

$$T_i^{la} = \frac{K_i |w_g|}{f_i^C} f_b^C. \qquad (4.21)$$

3. *Transmission of the model parameters:* The local models aggregated by BS i are then broadcast to other BSs as transactions. The time cost is related to the number of blockchain nodes and the transmission efficiency. Since other BSs also help to transmit the transaction in the broadcast process, the time function is related to $log_2 M$, where M is the size of the BS network. The required time cost is

$$T_i^{pt} = \xi log_2 M \frac{K_i |w_g|}{R_i^U}, \qquad (4.22)$$

where ξ is a factor of the transmission time cost that can be obtained from the historical running records of the transmission process.

4. *Block validation:* The block producer BS collects the transactions and packs them into a block. The block is then broadcast to other producer BSs and validated by them. Thus, the time cost is

$$T_{bp}^{bv} = \xi log_2 M_p \frac{S_B}{R_i^D} + \max_i \frac{S_B f^v}{f_i s}, \qquad (4.23)$$

where M_p is the number of block producers and S_B is the size of a block.

Note that, in the aggregation phase, the size of the model parameters $|w_g|$ is small and the computing capability f_i is high. Thus, compared to other phases, the time for aggregation is very short, such that it can be neglected. Based on the above analysis, the time cost for one iteration is denoted as

$$T = \max_i \left\{ \frac{\sum_{j=1}^{K_i} b_j D_{DT_j}}{f_i^C} f^C \right\} + \max_i \left\{ \xi log_2 M \frac{K_i |w_g|}{R_i^U} \right\}$$

$$+ \xi log_2 M_p \frac{S_B}{R_i^D} + \max_i \frac{S_B f^v}{f_i s}. \tag{4.24}$$

In the 6G network, the growth of the user scale, the ultra-low latency requirement of communication, and the dynamic network status make the reduction of the time cost of model training an important issue in various applications. Since accuracy and latency are the two main metrics for evaluating the decision-making abilities of digital twins in our proposed scheme, we consider the edge association problem to find the trade-off between learning accuracy and the time cost of the learning process. Due to the dynamic computing and communication capabilities of various BSs, the edge association of digital twins—that is, how to allocate the digital twins of different end users to various BSs for training—is a key issue to be solved to minimize the total time cost. Moreover, increasing the training batchsize b_n of each digital twin DT_n can improve the learning accuracy. However, this will also increase the learning time cost to execute more computations. In addition, how to allocate the bandwidth resources to improve communication efficiency should be considered. In our edge association problem, we should carefully design these policies to minimize the total time cost of the proposed scheme. Thus, we formulate the optimization problem as the minimization of the time cost of federated learning for a given learning accuracy. To solve the problem, the association of digital twins, the batchsize of their training data, and the bandwidth allocation should be jointly considered according to the computing capability f_i^C and the channel state $h_{i,c}$. The optimization problem can be formulated as

$$\min_{K_i, b_n, \tau_{i,c}} \frac{1}{1 - \theta_G} T \tag{4.25}$$

$$\text{s.t.} \quad \theta_G \geq \theta_{th}, \theta_G, \theta_{th} \in (0, 1), \tag{4.25a}$$

$$\sum_{i=1}^{M} K_i = D, K_i \in \mathcal{N}, \tag{4.25b}$$

$$\sum_{i=1}^{M} \tau_{i,c} \leq 1, c \in C, \tag{4.25c}$$

$$b^{min} \leq b_n \leq b_n^{max}, \forall n \in \mathcal{N}. \tag{4.25d}$$

Constraint (4.25b) ensures that the sum of the number of associated digital twins does not exceed the size of the total dataset. Constraint (4.25c) guarantees that each subchannel can only be allocated to at most one BS. Constraint (4.25d) ensures the range of the training batchsize for each digital twin. Problem (4.25) is a combinational problem. Since there are several products of variables in the objective function and the time cost of each BS is also affected by the resource states of other BSs, problem (4.25) is challenging to solve.

4.3.3 Multi-Agent DRL for Edge Association

Since the system states are only determined by network states in the current iteration and the allocation policies in the last iteration, we regard the problem as a Markov decision process and use a multi-agent DRL-based algorithm to solve it.

The proposed multi-agent reinforcement learning framework is depicted in Fig. 4.7. In our proposed system, each BS is regarded as a DRL agent. The environment consists of BSs and the digital twins of the end users. Our multi-agent DRL framework consists of multiple agents, a common environment, the system state S, the action $\mathcal{A}$, and the reward function $\mathcal{R}$, which are described below.

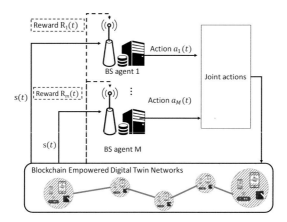

Fig. 4.7 Multi-agent DRL for edge association

- *State space:* The state of the environment is composed of the computing capabilities f^C of the BSs, the number of digital twins K_i on each BS i, the training data size of each digital twin D_n, and the channel state $h_{i,c}$. The states the of multiple agents are denoted as $s(t) = (f^C, K, D, h)$, where each dimension is a state vector that contains the states for all the agents.
- *Action space:* The actions of BS i in our system consist of the digital twin allocation K_i, the training data batchsizes for its digital twins b_i, and the bandwidth allocation τ_i. Thus, the actions are denoted as $a_i(t) = (K_i, b_i, \tau_i)$. BS agent i makes new action decisions $a_i(t)$ at the beginning of iteration t based on system state $s(t)$. The system action is $a(t) = (a_1, ..., a_i, ..., a_m)$.
- *Reward:* We define the reward function of BS i according to its time cost T_i based on Eq. (4.24):

$$\mathcal{R}_i(s(t), a_i(t)) = -T_i(t). \tag{4.26}$$

The reward vector of all the agents is $\mathbf{R} = (\mathcal{R}_1, ..., \mathcal{R}_m)$. According to Eq. (4.25), the total time cost T is decided by the maximum time cost of the agents $\max\{T_1, T_2, ..., T_m\}$. Each DRL agent in our scheme thus shares the same reward function. In the training process, the BS agents adjust their actions to maximize the reward function, that is, to minimize the system time cost in each iteration.

The learning process of BS i is to find the best policy that maps its states to its actions, denoted as $\boldsymbol{a}_i = \pi_i(s)$, where $\boldsymbol{a}_i$ is the action to be taken by BS i for the whole system state s. The objective is to maximize the expected reward, that is,

$$\mathcal{R}_t = \sum_i \gamma \mathcal{R}_i(s(t), \boldsymbol{a}_i(t)), \tag{4.27}$$

where γ is the discount rate, $0 \le \gamma \le 1$. In the conventional DRL framework, it is hard for an agent to obtain the states of others. In our DTWN, the states of the digital twins and BSs are recorded in the blockchain. A BS can retrieve records from the blockchain to obtain the system states and actions of other agents in the training process. We use $\pi = [\pi_1, \pi_2, ..., \pi_n]$ to denote the policies of the n agents, whose parameters are denoted as $\theta = [\theta_1, \theta_2, ..., \theta_n]$. Thus we have the following policy gradient for agent i:

$$\begin{aligned} \nabla_{\theta_i} J(\pi_i) = E_{\{\theta_i\}, a \sim D}[\nabla_{\theta_i} \pi(a_i|o_i)\cdot \\ \nabla_{a_i} Q_i^\pi(\{\theta_i\}, a_1, ..., a_n)|_{a_i = \pi_i(o_i)}], \end{aligned} \tag{4.28}$$

where $\{\theta_i\}$ is the observation of agent i, that is, the state of each agent. In our scheme, since the placement of digital twins requires global coordination, we consider that all the agents share the same system state through information exchange between the servers. Agent i determines its action $\boldsymbol{a}_i$ through its actor deep neural network (DNN) $\pi(s_t|\theta_\pi)$, denoted as

$$a_i(t) = \pi_i(s_t|\theta_{\pi_i}) + \mathfrak{N}, \tag{4.29}$$

where $\mathfrak{N}$ is the random noise for generating a new action. The actor DNN is trained as

$$\theta_\pi = \theta_\pi + \alpha_\pi \cdot \mathbb{E}[\nabla_{a_i} Q(s_t, a_1, ..., a_i|\theta_Q)|_{a_i = \pi(s_t|\theta_\pi)} \cdot \nabla_{\theta_\pi} \pi(s_t)], \tag{4.30}$$

where α_π is the learning rate of the actor DNN.

The critic DNN of agent i is trained as

$$\theta_{Q_i} = \theta_{Q_i} + \alpha_{Q_i} \cdot \mathbb{E}[2(y_t - Q(s_t, \boldsymbol{a}_i|\theta_{Q_i})) \cdot \nabla Q(s_t, a_1, ..., a_i)], \tag{4.31}$$

where α_{Q_i} is the learning rate, y_t is the target value, and $(\boldsymbol{a}_1, ..., \boldsymbol{a}_i)$ constitutes the actions of the agents in our system.

In the proposed algorithm, all the actor networks and critic networks are initialized randomly as the initial training parameters. Then the replay memory is initialized to store the experiential samples in the training process. In each episode, the agent selects its action towards its current observation state and obtains the reward for its current action. Then the new observation of the system state is obtained. The experience tuple (s_t, a_i, r_t, s_{t+1}) is then stored in the replay buffer. Finally, the agents train their critic network and actor network by sampling records from the replay buffer.

Chapter 5
Blockchain for Digital Twin

Abstract Security and privacy are critical issues in digital twin edge networks. Blockchain, as a tamper-proof distributed database, is a promising solution for protecting data and edge resource sharing in digital twin edge networks. In this chapter, we first introduce the architecture of the blockchain-empowered digital twin edge network, to show the integration angles of the blockchain and digital twin techniques. Then, we show the block generation and consensus process in the developed blockchain-empowered digital twin edge network architecture.

5.1 Blockchain-Empowered Digital Twin

Blockchain is a chain structure of data blocks arranged in chronological order that is essentially a tamper-proof distributed database that uses cryptography to ensure the security of each transaction in a decentralized manner. A blockchain is composed of peer-to-peer networks, distributed storage, consensus mechanisms, cryptography, and smart contracts. Therefore, a blockchain has the advantages of decentralization, tamper resistance, anonymity, public verifiability, and traceability [61, 62]. Integrating blockchain and digital twin provides security guarantee, trusted traceability, accessibility, and immutability of transactions in digital twin edge networks. Specifically, building virtual twins and continuously updating twin models require core data that contain private user information, such as production parameters and users' personal information in industrial manufacturing and personal health data in healthcare. Therefore, the physical and virtual synchronization during virtual twin construction and maintenance needs to be recorded as transactions in the blockchain. This means the core data and edge resources can be governed by the blockchain in a decentralized and secure manner. In addition, virtual twins can simulate the behavioural features of physical components and generate virtual data. After the blockchain stores the generated virtual data in its distributed ledger, these virtual data become digital assets and their ownership can be proved. To achieve a secure and reliable digital

61

Y. Zhang, *Digital Twin*, Simula SpringerBriefs on Computing 16,
https://doi.org/10.1007/978-3-031-51819-5_5

twin edge network, the following blockchain-related security performances should be satisfied.

- *Security and trust:* In the developed blockchain-empowered digital twin edge network, the transactions are audited and verified by a set of verifiers by utilizing consensus algorithms [63], unlike traditional transaction management, which depends on a central infrastructure. Thus, the blockchain can guarantee security and trust for the transactions in a digital twin edge network in a decentralized manner without a trusted intermediary.
- *Unforgeability and immutability:* The decentralized authentication of transactions in the blockchain-empowered digital twin edge network ensures that no attacker can pose as a user to corrupt the blockchain. In addition, verifiers who execute the consensus algorithms are reluctant to misbehave or collude with each other, since all the verifiers' identities are revealed to the users in a digital twin edge network and would be scrutinized for any misconduct. Furthermore, an attacker cannot modify the audited and stored transactions in the blockchain, since each block is embedded with the hash value of its previous block, which ensures immutability [64].
- *Transparency and privacy protection:* In the blockchain-empowered digital twin edge network, all the kinds of information recorded in the blockchain are transparent and openly accessible to all participants. Moreover, end users can change their identity (i.e. public key) after each transaction in the blockchain to protect their identity privacy.
- *Scalability and interoperability:* Digital twins are digital replicas of physical entities, enabling close monitoring, real-time interactions, and reliable communications between digital space and physical systems. They provide rich information to reflect the states of physical entities, to optimize the running of physical systems [65]. Therefore, the blockchain needs to provide scalability and interoperability for various digital twin services. The scalability of blockchain provides end users simultaneous access to the digital twin edge network. Meanwhile, the interoperability allows different digital replicas of physical entities to interact with each other seamlessly.

The architecture of a blockchain-empowered digital twin edge network is shown in Fig. 5.1. In the physical plane, physical objects share information with each other. Various wireless/wired devices, such as sensors, radio frequency identification devices, actuators, controllers, and other tags can connect with IoT gateways, Wi-Fi access points, and base stations (BSs) supporting the communications between physical objects. In the virtual plane, the digital twin edge servers provide the necessary computing capability to generate virtual twins of the physical objects, as well as model the channel conditions and data transmission among the physical objects. Moreover, a physical object realizes information transmission with a virtual twin through wireless communication technologies and shares the data in real time and accepts feedback from the virtual twin. In the blockchain plane, BSs are distributed in a specific area to work as verifiers. Specifically, if data or network resources are successfully shared between a requester and a provider in both the physical and virtual planes, the requester should create a transaction record and send it to the nearest BS. The BSs collect and manage local transaction records. The transaction records are structured into blocks after the consensus process among the BSs is completed, and then stored permanently in each BS. The detailed processes are as follows.

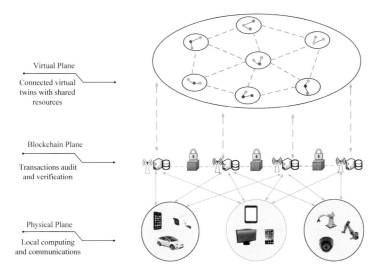

Fig. 5.1 The architecture of a blockchain-empowered digital twin edge network

- *System initialization:* For privacy protection, each device needs to register a legitimate identity in the initialization stage. In the blockchain-empowered digital twin edge network, an elliptic curve digital signature algorithm and asymmetric cryptography are used for initialization. A device can obtain a legitimate identity after its identity has been authenticated. The identity includes a public key, a private key, and the corresponding certificate.
- *Role selection for devices:* Devices in both the physical and virtual planes choose their roles (i.e. data or network resource requester and provider) according to their current available resources. Devices with surplus resources can become providers to provide services for requesters.
- *Transactions:* The existence of connections in the digital twin edge network poses new challenges for security and privacy. Therefore, communications in the physical and virtual planes can be considered as transactions, which are recorded into the blockchain to achieve security and preserve privacy. Additionally, the blockchain can store the synchronization data between the physical and virtual planes, so that the data are secure and private. In addition, the blockchain enables edge servers belonging to different stakeholders to cooperatively process computation tasks without a trusted third party.
- *Building blocks:* BSs collect all the transaction records within a certain period and then encrypt and digitally sign them to guarantee their authenticity and accuracy.

The transaction records are structured into blocks, and each block contains a cryptographic hash of the prior block. To verify the correctness of a new block, the consensus algorithm is used. In the proof-based consensus process, one of the BSs is selected as the leader for creating the new block. Because of broadcasts, each BS can access the entire transaction record and has the opportunity to be the leader.

- *The consensus process:* The leader broadcasts the created block to the other BSs for verification and audit. All the BSs audit the correctness of the created block and broadcast their audit results. The leader then analyses the audit results and, if necessary, sends the block back to the BSs for another audit.

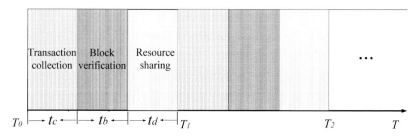

Fig. 5.2 The time sequence of the transaction confirmation procedure

5.2 Block Generation and Consensus for Digital Twin

In the developed blockchain-empowered digital twin edge network, the distinct characteristics of the blockchain introduce unique challenges for data and edge resource sharing. Specifically, Fig. 5.2 illustrates the time sequence of T rounds of the transaction confirmation procedure. In each round, the transaction confirmation time is divided into three time slots: a transaction collection slot, a block verification slot, and a resource sharing slot. For example, in the T_1 round, this consists of the slot t_c for transaction collection, the slot t_b for block verification, and the slot t_d for resource sharing among the edge devices. The requester can obtain the required resource from the provider only at the end of round T_1's transaction confirmation procedure, that is, when the block is appended to the blockchain. In practice, it could take a long time to successfully finish the transaction confirmation procedure, since both transaction collection and block verification are executed in a dynamic and stochastic wireless transmission environment. Transactions of edge devices in congested areas might not be successfully transmitted to the verifiers for verification in the transaction collection phase, which will lead to failure of the transaction confirmation procedure. On the other hand, the typical Nakamoto consensus protocol provides proof of work (PoW) [66], where the verifiers compete to solve a computationally difficult cryptopuzzle. The fastest verifier that solves the cryptopuzzle will append its block to the blockchain. Nevertheless, the cryptopuzzle solving–based PoW consensus protocol consumes a large amount of computational and energy resources, which is not useful for resource-constrained edge devices, since they cannot undertake heavy computations. In addition, the audit and verification of the block among the verifiers can encounter impediments due to traffic congestion in the network. Therefore, the edge devices could suffer from long waiting times in resource sharing. A carefully designed transaction confirmation procedure is thus necessary for secure and privacy-protected resource sharing among edge devices while allowing for their resource sharing efficiency.

To enable edge devices to obtain the required resources in time, we present a block generation and consensus process in the developed blockchain-empowered digital twin edge network. The process is based on a relay-assisted transaction relaying scheme that facilitates transaction collection in congested areas, and a lightweight block verification scheme based on delegated proof of stake (DPoS) that is utilized to reduce the resource consumption of the verifiers during block verification.

- In the transaction collection phase, the local verifiers periodically collect transactions, verify the integrity and correctness of the transactions by validating their signature, and then process a number of validated transactions into a block.
- In the block verification phase, the local verifiers that would like to add a block to the blockchain send consensus requests to a verification set, which consists of a set of preselected verifiers, and executes block verification and audit by using a proof-based consensus protocol.

We develop two new schemes for transaction collection and block verification: (I) a relay-assisted transaction relaying scheme and (II) a DPoS-based lightweight block verification scheme. The work procedure for the developed schemes is shown in Fig. 5.3 and is illustrated in the subsequent discussion.

5.2.1 Blockchain Model

To enhance the security and reliability of digital twins from untrusted end users, the BSs act as blockchain nodes and maintain the running of the permissioned blockchain. The digital twins are stored in the blockchain and their data are updated as the states of the corresponding users change. The local models of the BS, also stored in the blockchain, can be verified by other BSs to ensure their quality. Thus, there are three types of records, namely, digital twin model records, digital twin data records, and training model records.

The overall digital twin blockchain scheme is shown in Fig. 5.4. The BSs first train the local training models on their own data and then upload the trained models to the MBS. The trained models are also recorded as blockchain transactions and are broadcast to the other BSs for verification. The other BSs collect the transactions and pack them into blocks. The consensus process is executed to verify the transactions

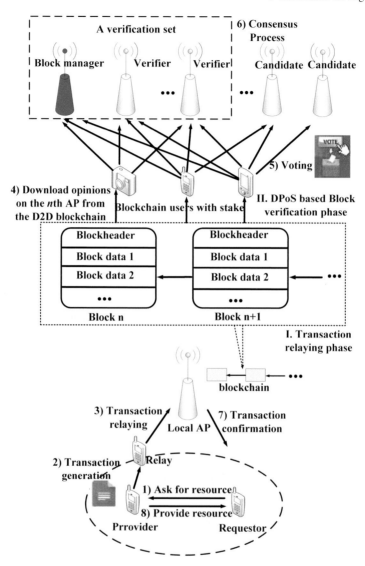

Fig. 5.3 Work procedure for transaction relaying and DPoS-based block verification

in blocks. Our consensus process is executed based on the DPoS protocol, where the

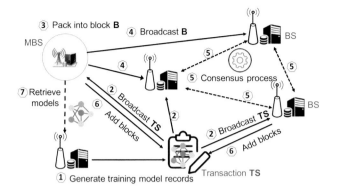

Fig. 5.4 The blockchain scheme for federated learning

stakes are the training coins. The initial training coins are allocated to BS i according to its data from digital twins, denoted as

$$S_i = \frac{\sum_{j=1}^{K_i} D_{DT_j}}{\sum_{k=1}^{M} D_k} S_{ini}, \tag{5.1}$$

where S_{ini} is an initial value and K_i is the number of digital twins associated with BS i.

The coins of each BS are then adjusted according to their performance in the training process. If the trained model of a BS passes the verification of the other BSs and the MBS, the coins will be awarded to the BS. Otherwise, the BS will receive no pay for its training work. A number of BSs are elected as block producers by all the BSs. In the voting process, all the BSs vote for the candidate BSs by using their own training coins. The elected BSs take turns to pack the transactions in a time interval T into a block B and broadcast the block B to other producers for verification.

In our proposed scheme, we leverage blockchain to verify the local models before embedding them into the global model. Due to high resource consumption required for block verification, the interval T should be set to multiple times the local training period; that is, the BSs execute multiple local training iterations before transmitting the local models to the MBS for global aggregation.

5.2.2 Relay-Assisted Transaction Relaying Scheme

As shown in Fig. 5.3, a requester that requires resources first sends a request to a nearby provider (Step 1). Then the provider generates a transaction (Step 2) and broadcasts its transaction for verification. A provider that would like to verify its transaction in a congested area first sends a transaction relaying request. To facilitate peer discovery and reduce interference, the local verifier coordinates the transaction relaying link establishment; that is, the verifier selects a device near the provider as a relay device to assist in transaction relaying and reuses the channels of end users located far away in a different area (Step 3). We consider the time division multiple access technique in the transaction relaying transmission. However, the existence of a neighbour might not imply the stable establishment of the transaction relaying link, since the neighbour might not be willing to participate in transaction relaying due to the associated overhead, such as energy and bandwidth consumption. Thus, the local verifier needs to pay the relay devices a relay fee to motivate them to participate in transaction relaying.

5.2.3 DPoS-Based Lightweight Block Verification Scheme

DPoS has been demonstrated as a high-efficiency consensus protocol with moderate cost in which a part of the delegates (i.e. verifiers) are selected based on their stakes to perform the consensus process. Here, the stake is the accumulated time during which a delegate possesses its assets before using them to generate a new block. DPoS has been used in real scenarios, such as enterprise operation systems, BitShares, and the Internet of Vehicles. It is reasonable to consider that the DPoS can be utilized in the digital twin edge network and to develop a DPoS-based lightweight block verification scheme. Unlike computation-intensive PoW, the designed DPoS-based lightweight block verification scheme can leverage the stake of the verifiers as a mining resource to generate a block. The more stakes a verifier has, the higher its probability of finding a solution to generate a block. In addition, the verification and audit of the generated block are executed by only some of the preselected verifiers, thus keeping the computational complexity reasonably low. As shown in Fig. 5.3, the main steps in the DPoS consensus protocol in our lightweight block verification scheme involve verifying candidate generation, the verifier selection, the consensus process, and the transaction fee payment.

5.2.3.1 Verifying Candidate Generation

A verifier that wants to be a verifier first submits a deposit of stake to an account under public supervision. This deposit will be confiscated if the verifier behaves badly during a consensus process, for example, if it fails to generate a block in its turn or if it generates false block verification results.

5.2.3.2 Verifier Selection

The blockchain users, that is, the end users possessing stakes, download the opinions of the candidates from the blockchain (Step 4) and vote for their preferred candidates according to some criteria, for example, voting for candidates that can generate and verify a block quickly (Step 5). A blockchain user can vote for more than one candidate and can also persuade others to vote for their favourite candidates. The top k candidates with the most votes are selected to form a verification set, where k is an odd integer, such as 21 in enterprise operation systems. The k verifiers all take turns acting as the block manager during k block verification subslots.

5.2.3.3 Consensus Process

In each block verification subslot, the block manager carries out block management in its own consensus process round (Step 6). Specifically, the block manager first broadcasts the unverified block to other verifiers for verification and audit. Then, each verifier locally verifies the signature of each transaction in the block and replies to other audit results with its signature. Following the reception of the audit results, each verifier compares its audit result with those of the other verifiers and sends a commit message to the block manager. Considering Byzantine fault tolerance consensus conditions, the block manager sends the current audited block to all the verifiers and the local verifiers for storage (Step 7), providing it receives a commit message from more than two-thirds of the verifiers. Finally, the provider provides the required resources to the requestor (Step 8).

5.2.4 Conclusion

We first presented the architecture of a blockchain-empowered digital twin edge network that consists of a virtual plane, a blockchain plane, and a physical plane. Then, we illustrated the processes of the developed architecture and showed the integration angles of the blockchain and the digital twin edge network. Furthermore, we presented the block generation and consensus process in the developed blockchain-empowered digital twin edge network architecture.

Chapter 6
Digital Twin for 6G Networks

Abstract Digital twin is a technology that has the potential to help sixth-generation (6G) networks to realize digitization. In this chapter, we first introduce the combination of digital twin and 6G and then discuss two key use cases in terms of reconfigurable intelligent surfaces and digital twin and digital twins for stochastic offloading.

6.1 Integration of Digital Twin and Sixth-Generation (6G) Networks

To meet the ever-increasing demands of user traffic, fifth-generation (5G) networks integrate several novel network architectures, such as edge computing, software-defined networking, network function virtualization, and ultra-dense heterogeneous networks, to realize performance improvements for peak rates, transmission latency, network energy efficiency, and other indicators. However, the rapid proliferation and breakneck expansion of 5G wireless services also pose new challenges on transmission data rates, ubiquitous coverage, reliability, and network intelligence [67]. These challenges are spurring activities focused on defining the next-generation 6G wireless networks. Compared with 5G, 6G networks are envisioned to achieve the superior performance in the following areas [68, 69].

- *Peak data rate*: The peak data rate is the highest data rate under ideal channel conditions where all available radio resources are completely assigned to a single mobile device. Driven by both user demand and technological advances such as terahertz communications, peak data rates are expected to reach up to 1 Tbps, 10 times that of 5G.
- *Latency*: Latency can be distinguished as the user plane and control plane latency. The minimum latency requirement for the user plane is 1–4 ms. This value is envisioned to be further reduced in 6G to 100 μs or even 10 μs. The minimum

© The Author(s) 2024

Y. Zhang, *Digital Twin*, Simula SpringerBriefs on Computing 16,
https://doi.org/10.1007/978-3-031-51819-5_6

latency for the control plane should be 10 ms in 5G and is also expected to be remarkably improved in 6G.

- *Mobility*: The highest mobility supported by 5G is 500 km/h. In 6G, the maximal speed of 1,000 km/h is targeted to meet the requirements of commercial airline systems.
- *Connection density*: The minimum number of devices with a relaxed quality of service in 5G is 10^6/km^2. In 6G, the connection density is envisioned to be further improved by 10 times, to 10^7/km^2.
- *Energy efficiency*: Energy efficiency is an important metric to enable cost-efficient wireless networks for green communications. In 6G, network energy efficiency is expected to increase 10 to 100 times compared to that in 5G.
- *Signal bandwidth*: The requirement for bandwidth in 5G is at least 100 MHz, and 6G will support up to 1 GHz for operations in higher frequency bands, and even higher in terahertz communications.

Beyond imposing new performance metrics, emerging trends that include new services and the recent revolutions in artificial intelligence (AI), computing, and sensing will redefine 6G. Digital twin, as one of the emerging technologies for next-generation network digitalization, can pave the way for the creation of future digital 6G by transforming and precisely mapping physical networks to digital networks with virtual twins. Digital twin will provide three main benefits for 6G. First, digital twin can provide a comprehensive and accurate network analysis for 6G with increasingly accurate and synchronous network updates. Second, digital twin can build a virtual twin layer between the physical entities and user applications. This can establish a bridge between the bottom network and the top application with better cross-layer interaction and timely user experience feedback. Third, digital twin–enabled 6G can utilize AI algorithms to adjust network schedules, such as task offloading, resource allocation, and network management. Thus, digital twin is an essential technique for 6G in terms of supporting network automation and intelligence.

6.2 Potential Use Cases

Several works have explored utilizing digital twins to enhance the performance of next-generation communication networks. In [70], the authors proposed digital twin–enabled 6G to enable network scalability and reliability. The authors in [71] analysed the potential of digital twin for next-generation communication networks in terms of radio access, channel emulation, and network optimization. These works discussed how digital twin could be a powerful tool to fulfil the potential of 6G. Next, we present three detailed use cases of the combination of digital twin and 6G.

- *Reconfigurable intelligent surface (RIS) technology and digital twin*: With the dense deployment of edge servers, there will be increasing data transmission requirements in the next-generation networks, which will aggravate network interference and increase transmission delays. Current massive multiple input, multiple output and millimetre wave technologies can increase wireless communication data rates, but these can incur high hardware costs and complicated signal processing issues. RIS is a new technology for 6G that can enhance spectral efficiency and suppress interference in wireless communications by adaptively configuring massive low-cost passive reflecting elements. However, to improve wireless transmission rates, RIS requires both the amplitude and phase of passive reflecting elements to be adjusted to facilitate an enhanced signal propagation environment. Since virtual twins can record the real-time states of physical objects, monitor the dynamic changes of wireless networks, and carry out optimization and prediction to improve the performance of the physical system, RIS can utilize digital twin to extract the key features of RIS components, such as the number of RIS elements, the phase and amplitude of the reflecting elements, and the mobile devices served by each RIS element. With the extracted information, digital twin can assist in RIS to adjust the wireless propagation environment to improve the signal-to-noise ratio and decrease the probability of outages.

- *Edge association and digital twin*: The huge number of connected devices and the heterogeneous network structure of 6G pose great challenges for constructing digital twins in each network's infrastructure. A possible solution for this issue is to select a subset of base stations as the digital twin servers to maintain the digital twins at reduced time cost and energy consumption, instead of maintaining digital twins at every base station (BS). To achieve this, the edge association problem must be addressed. The objective of edge association is to minimize the average system latency while providing delay-guaranteed service for each user. According to the running phases of digital twins, edge association consists of two subproblems: the digital twin placement problem and the digital twin migration problem. The digital twin placement problem involves how to choose the optimized subset of BSs as digital twin servers. The migration of digital twins problem involves how to allocate network resources to ensure relatively low transmission overhead and communication latency in the process of digital twin migration.

- *Cellular vehicle to everything (C-V2X) and digital twin*: The rapid development of wireless communications and C-V2X has facilitated the wide use of smart vehicles and enriched many intelligent transportation system applications, such as smart navigation, road condition recognition, high-precision real-time mapping, forward collision warning, and driving assistance. However, due to the high mobility of vehicles, it is difficult to test C-V2X functionalities and performance for typical V2X use cases. Digital twin can provide a high-fidelity digital mirror of C-V2X systems throughout their entire life cycle [72]. By using digital twin mapping, the predicted state of automatic driving vehicles can be realized based on a virtual simulation test environment. Based on the prediction information of digital twin, driving behaviours and emergency events can be more actually determined and quickly perceived.

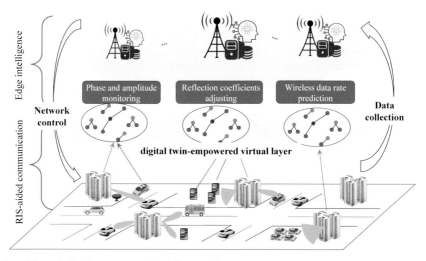

Fig. 6.1 Digital twin–empowered RIS framework

6.3 Digital Twin for RIS

To support emerging applications, 6G networks deploy computation/storage capabilities at BSs to avoid long transmission delays from mobile devices to cloud servers. However, while this shortens the distance and delay to access cloud server resources, it does not improve the wireless propagation environment. The recently proposed RIS technology can enhance spectral efficiency and suppress interference by adjusting both the amplitude and phase of passive reflecting elements. Digital twin can assist in RIS to intelligently adjust passive reflecting elements.

6.3.1 System Model

To clearly illustrate the combination of digital twin and RIS, we present a hierarchical digital twin–empowered RIS framework, as shown in Fig. 6.1. In this framework, edge resources can alleviate the heavy computational pressure of mobile devices, and edge servers can reduce task processing latency due to their proximity to mobile devices. RIS can enhance the quality of wireless communication links in the process of task offloading by intelligently altering the radio propagation environment.

The proposed framework consists of two layers: an RIS-aided communication layer and a digital twin–empowered virtual layer. In the RIS-aided communication layer, RIS elements are distributively installed on the surface of building facades, to improve propagation conditions and increase the quality of wireless communications. The digital twin–empowered virtual layer is constructed by diverse distributed edge servers. With edge resources and AI algorithms, virtual twins can construct a real-time mirror of the physical network to enable intelligent policy design, quality of service requirements, resource management, and network topology monitoring. This is a general framework that can improve the communication and computational performance in many scenarios, including cellular, vehicular, and unmanned aerial vehicle networks.

6.3.2 Computation Offloading in Digital Twin–Aided RIS

To elaborate on how digital twin assists in RIS coefficient adjustment, in this section, we present a case study that focuses on RIS-aided offloading. We consider a network of digital twin–aided RIS offloading that consists of a physical network entities layer and a digital twin–empowered virtual layer. The physical network entities layer contains three types of physical entities: base stations, RISs, and mobile devices. Since digital twin mirrors a physical entity, the digital twin–empowered virtual layer also contains three types of virtual models. The first type of virtual model involves the BSs. We consider that each physical BS has multiple antennas and an edge server for providing edge computing via wireless communications. The virtual model of a BS with edge intelligence and can thus predict current available communication, computing, and caching resources and monitor current wireless links to construct the current network topology. The second type of virtual model involves RIS, including the number of RIS elements and the phase and amplitude of reflecting elements. The key function of this virtual model is to adjust the RIS coefficients. The third type of virtual model involves mobile devices. This type of virtual model mainly records the size of the collected data, the current locations of the mobile devices, and the latency or computational resource requirements of on-device applications.

Task offloading aims to offload the computation-intensive tasks of mobile devices to nearby distributed BSs for processing. The virtual model of each mobile device records the computation-intensive task as (d_k, c_k), where d_k is the data size of task k and c_k is the required computation resource for the computing unit bit. The virtual model needs to determine what part of the task should be processed locally and how much should be offloaded to the edge server to process. We define this as the offloading ratio (i.e. x_k). RIS offloading utilizes RIS to assist in task offloading for a higher wireless communication rate. Different from traditional wireless transmission links, which only include direct device–BS links, the wireless transmission link in RIS offloading includes both of device–BS links and reflected device–RIS–BS links. For the device–BS link, the virtual model of the BS records its channel vector, that is, $\mathbf{h}_k^d$. The reflected device–RIS–BS link contains three components: the device–RIS link, the RIS reflection with phase shifts, and the RIS–BS link. The virtual RIS model records the channel vectors of the device–RIS link and RIS–BS link as $\mathbf{h}^r$ and $\mathbf{h}^H$, respectively. The RIS reflection coefficients are denoted as $\Theta = \mathrm{diag}(\beta_1 e^{j\theta_1}, \beta_2 e^{j\theta_2}, ..., \beta_N e^{j\theta_N})$, where β_n and θ_n are the amplitude and phase shift of the nth RIS element, respectively. The effective channel gain can be expressed as

$$\mathbf{g}_k = \mathbf{h}_k^d + \mathbf{h}^H \Theta \mathbf{h}_k^r. \tag{6.1}$$

Based on channel gain, the maximum achievable wireless transmission data rate can be obtained by

$$R_k = B \log_2 \left(1 + \frac{p_k |\mathbf{w}_k^H \mathbf{g}_k|^2}{\sum_{j=1,j\neq i}^K p_j |\mathbf{w}_i^H \mathbf{g}_k|^2 + \sigma^2} \right), \tag{6.2}$$

where B is the system's bandwidth. The virtual RIS model should properly adjust the reflection coefficients to improve the wireless communication rate.

The task execution latency is determined by the local computation and task offloading. The latency of local computation is mainly related to the computational capability of each mobile device (i.e. f_k^l). The latency of task offloading involves the task transmission time and edge computation time. Since the two parts are executed in parallel, the total task execution latency is equal to the maximal value of the two processes. To minimize the total task execution latency, the RIS configuration offloading ratio and computation resource must be jointly optimized. The RIS offloading problem can be formulated as

$$\min_{x,f,\beta,\theta} \sum_{k \in \mathcal{K}} \max\{ d_k c_k \frac{1-x_k}{f_k^l}, d_k c_k \frac{x_k}{f_k^s} + d_k \frac{x_k}{R_k(\mathbf{g}_k)} \}$$

$$\text{s.t.} \sum_{k \in \mathcal{K}} f_k^s \leqslant F^s, \ 0 \leqslant f_k^s \leqslant F^s, k \in \mathcal{K}, \tag{6.3a}$$

$$x_k, \beta_n \in [0,1], k \in \mathcal{K}, n \in \mathcal{N}, \tag{6.3b}$$

$$0 \leqslant \theta_n \leqslant 2\pi, n \in \mathcal{N}, \tag{6.3c}$$

where f_k^s and F^s are the computation resource that the BS allocates to task k and the total computation resource of the BS. Constraint (6.3a) is the computation resource allocation constraint. Constraints (6.3b) and (6.3c) are the value ranges of the offloading ratio, amplitude and phase shift variables, respectively. Since the digital twin–empowered virtual layer has AI ability, we can use AI, such as deep reinforcement learning (DRL), to solve the complex optimization problem. We first reformulate the above optimization problem as DRL with a system state, action, and reward. The state has five components:

$$s(t) = \{d_k(t), c_k(t), f_k^l(t), F_s, \Theta(t)\}. \tag{6.4}$$

In the environment, the BS assembles the information as a state and sends it to the DRL agent. The action has four parts, which are the variables of the optimization problem:

$$a(t) = \{x_k(t), f_k^s(t), \beta_n(t), \theta_n(t)\}. \tag{6.5}$$

Based on the state and action, the agent can produce a reward $\mathcal{R}^{imm}(s(t), a(t))$ from the environment, where the reward is related to the objective function. In this scenario, the total task execution latency can be regarded as the reward function.

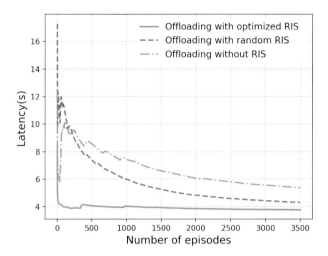

Fig. 6.2 Cumulative task execution latency under different schemes

Based on the state, action, and reward, we exploit asynchronous actor–critic DRL to solve the formulated problem [73]. Asynchronous actor–critic DRL consists of a global agent and several local agents. The global agent accumulates all the parameters of the neural networks from the local agents. Each local agent has an actor neural network and a critic neural network. The actor neural network is for generating actions and the critic neural network is for evaluating the performance of the action generated by the actor neural network. At each training step, the parameter of the actor neural network is updated based on

$$\theta_\pi \leftarrow \theta_\pi + \alpha_\pi \sum_t \nabla_{\theta_\pi} \log \pi(s(t)|\theta_\pi)(\mathcal{R}^{imm}(s(t), a(t))$$
$$+ \delta v_{\theta_v}(s(t+1)) - v_{\theta_v}(s(t))),$$
$$(6.6)$$

where α_π is the learning rate of the actor network and $\pi(s(t)|\theta_\pi)$ is the output of the actor neural network. The parameter of the critic neural network is updated based on

$$\theta_v \leftarrow \theta_v + \alpha_v \sum_t \nabla_{\theta_v}(\mathcal{R}^{imm}(s(t), a(t)) + \delta v_{\theta_v}(s(t+1)) - v_{\theta_v}(s(t))))^2, \quad (6.7)$$

where α_v is the learning rate of the critic network.

Figure 6.2 shows the total task execution latency of computation offloading under different RIS configuration schemes. First, we can see that the proposed DRL-based computation offloading algorithm converges in all cases and the cumulative task execution latency reduces with the number of episodes. Further, the offloading latency with RIS aid is lower than the latency without RIS aid. The reason is

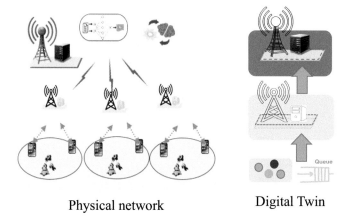

Physical network   Digital Twin

Fig. 6.3 Illustration of a digital twin network

that RIS offloading can achieve a higher transmission data rate, thus resulting in a lower transmission latency. In addition, the offloading latency with optimized RIS configuration is the lowest due the optimal adjustments of the RIS amplitude and phase shift.

6.4 Stochastic Computation Offloading

To improve task processing efficiency and prolong the battery lifetime of mobile devices, computation offloading is a promising approach that can offload the collected data and computation tasks to distributed BSs for processing. However, current research focusing on computation offloading assumes that each device executes a single computation task, without considering the randomness of task arrivals. Such an assumption in the designed policy cannot be applied to a network with a stochastic task arrival model. Since digital twin is a powerful technology that can monitor and analyse the dynamic changes of physical objects, in this section, we utilize digital twin to construct virtual models of the physical objects and solve the stochastic computation offloading problem considering dynamic changes of the task queue.

6.4.1 System Model

We consider a digital twin network consisting of a physical network and its digital twin. As shown in Fig. 6.3, the physical network has three major components: distributed mobile devices, small base stations (SBSs), and a macro base stations (MBS). Each device collects data from sensors and on-device applications, and the collected data must be analysed in real time. Since data analysis is computation intensive, devices with limited computation capability and battery power might not be able to conduct the data analysis in a timely manner. So, the devices must offload these tasks to edge servers for a high quality of computational experience. Digital twins contain the virtual models of the physical elements. Virtual models not only mirror the characteristics of the physical elements/system, but also make predictions, simulate the system, and can play a crucial role in policy design and resource allocation. In the network, digital twin can be utilized [74] to 1) construct the network topology of the physical network; 2) monitor network parameters and models, that is, dynamic changes of resources and stochastic task arrival processes, and 3) optimize offloading and resource allocation policy.

6.4.2 Stochastic Computation Offloading: Definition and Problem Formulation

Based on digital twin, the digital representation (i.e. virtual models) of the physical network (i.e. virtual world) is created. The virtual models here comprise the wireless network topology, the communication model between the devices and BSs, and the stochastic task queueing model.

(1) Network topology in the digital twin network

Digital twin first models the physical network as a graph $G = (\mathcal{U}, \mathcal{B}, \varepsilon)$, where $\mathcal{U} = \{u_1, .., u_N\}$ and $\mathcal{B} = \{b_0, b_1, ..., b_M\}$ are, respectively, the sets of devices and BSs (where b_0 is the index for the MBS, and the other values are the indexes for the SBSs). The term ε is the edge information, that is, for the connection between the devices and BSs.

Then, the digital twin uses a 3-tuple $DT_i(t)$ to characterize devices, that is, $DT_i(t) = \{p_{i,max}(t), l_i(t), f_i^l\}$, where $p_{i,max}(t)$ denotes the maximal transmission power in time slot t, $l_i(t)$ denotes the current location of u_i, and f_i^l denotes the computation resources of the local server. Similarly, the digital twin uses a 3-tuple $DT_j(t)$ to characterize the BSs, that is, $DT_j(t) = \{l_j(t), w_j, f_j^e\}$, where $l_j(t)$ denotes the current location of b_j, w_j denotes the bandwidth of b_j, and f_j^e denotes the computation resource.

The task offloading between the devices and BSs is facilitated through wireless communication. Here, we consider that devices communicate with the nearest BS for offloading. The wireless communication data rate between device u_i and SBS b_j can be expressed as

$$R_{ij}^s(t) = w_{ij}(t) \log(1 + \frac{p_i(t)h_{ij}^s(t)r_{ij}^s(t)^{-\alpha}}{\sigma^2 + I}), \qquad (6.8)$$

where $w_{ij}(t)$ ($w_{ij}(t) \leq w_j$) is the bandwidth that SBS b_j allocates to device u_i in time slot t, $h_{ij}^s(t)$ is the current channel gain, α is the path loss exponent, σ^2 is the noise power, $r_{ij}^s(t)$ is calculated based on the locations of $l_i(t)$ and $l_j(t)$, and I is the

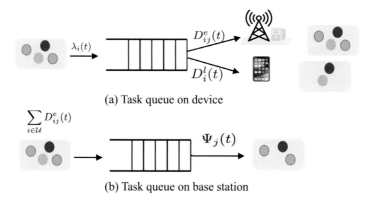

(a) Task queue on device

(b) Task queue on base station

Fig. 6.4 Stochastic computation offloading in digital twin network

interference from other SBSs. With the adoption of orthogonal frequency division multiple access, the interference of different devices in the coverage of the MBS is ignored. The wireless communication data rate between device u_i and the MBS is

$$R_{i0}^m(t) = w_{i0}(t) \log(1 + \frac{p_i(t)h_{i0}^m(t)r_{i0}^m(t)^{-\alpha}}{\sigma^2}), \qquad (6.9)$$

where $w_{i0}(t)$ $(w_{i0}(t) \leq w_0)$ is the channel bandwidth between device u_i and the MBS in time slot t, $h_{i0}^m(t)$ is the channel gain between device u_i and the MBS, and $r_{i0}^m(t)$ is the distance between device u_i and the MBS.

(2) Stochastic task queueing

At the beginning of time slot t, device u_i inputs the size of the computation task of $\lambda_i(t)$ (bits/slot) into the local dataset. We assume the $\lambda_i(t)$ values in different time slots are independent, and $\mathbb{E}[\lambda_i(t)] = \lambda$. Since device u_i has computation resources, it can execute part of the computation task locally. We consider the size of the computation task that is executed locally as $D_i^l(t)$. The size of the computation task offloaded to BS $b_j(j \in \mathcal{B})$ is $D_{ij}^e(t)$. The rest is stored in a local task buffer, as shown in Fig. 6.4(a). Assume the queue length of the local task buffer is $Q_i^l(t)$ and the queue length is dynamically updated with the following equation:

$$Q_i^l(t + 1) = \max\{Q_i^l(t) - \Psi_i(t), 0\} + \lambda_i(t), \qquad (6.10)$$

where $\Psi_i(t) = D_i^l(t) + D_{ij}^e(t)$ is the size of the computation task that leaves the task buffer of device u_i during time slot t.

Each edge server also has a task buffer to store the offloaded but not yet executed task. As shown in Fig. 6.4(b), the queue length is dynamically updated by

$$Q_j^e(t + 1) = \max\{Q_j^e(t) - \Psi_j(t), 0\} + \sum_{i \in \mathcal{U}} D_{ij}^e(t), \qquad (6.11)$$

where $\sum_{i \in \mathcal{U}} D^e_{ij}(t)$ is the amount of tasks offloaded from all the devices connected to BS j, and $\Psi_j(t)$ is the size of the computation tasks leaving the edge task buffer. According to the definition of stability in [75], the task queue is stable if all the computation tasks satisfy the following constraints:

$$\lim_{T \to \infty} \frac{1}{T} \sum_{t=0}^{T-1} \sum_{i \in \mathcal{U}} \mathbb{E}\{Q^l_i(t)\} < \infty, \tag{6.12a}$$

$$\lim_{T \to \infty} \frac{1}{T} \sum_{t=0}^{T-1} \sum_{j \in \mathcal{B}} \mathbb{E}\{Q^e_j(t)\} < \infty. \tag{6.12b}$$

(3) Task offloading in the digital twin network

Let $f^l_i(t)$ be the computation resource of device u_i during time slot t and let c denote the required computation resource for executing one bit of a computation task. Thus, the size of computation tasks executed locally will be

$$D^l_i(t) = \frac{\tau f^l_i(t)}{c}, \tag{6.13}$$

where τ is the duration of the time slot. The energy consumption of a unit of computation resource is $\varsigma(f^l_i)^2$, where ς is the effective switched capacitance, depending on the chip architecture. The local energy consumption for computing task $D^l_i(t)$ can be defined as

$$E^l_i(t) = \varsigma \tau f^l_i(t)^3. \tag{6.14}$$

Devices offload their tasks to BSs via wireless communication. Since the devices are associated with different BSs, the offloaded tasks of device u_i during time slot t can be expressed as

$$D^e_{ij}(t) = \begin{cases} R^s_{ij}(t)\tau & j \in \mathcal{B}/\{b_0\}, \\ R^m_{i0}(t)\tau & j = b_0. \end{cases} \tag{6.15}$$

The energy consumption in this case has three parts: the energy consumption for uplink offloading, the energy consumption for computation, and the energy consumption for downlink feedback. The third quantity is generally ignored due to its small data size. Thus, the energy consumption for executing task $D^e_{ij}(t)$ on BS b_j can be expressed as

$$E^e_{ij}(t) = p_i(t)\tau + \frac{D^e_{ij}(t) * c}{f^e_{ij}(t)} * \varepsilon, \tag{6.16}$$

where $f^e_{ij}(t)$ is the computation resource that b_j allocates to device u_i in time slot t, and ε is the energy consumption for unit computation on edge servers.

The total energy consumption is the combination of local energy consumption, edge server energy consumption, and the transmission energy consumption for computation offloading. Therefore, the total energy consumption can be expressed as

$$E^{tol}(t) = \sum_{i \in \mathcal{U}} E_i^l(t) + \sum_{i \in \mathcal{U}} \sum_{j \in \mathcal{B}} E_{ij}^e(t). \tag{6.17}$$

(4) Stochastic offloading problem

Based on the total energy consumption, we can define network efficiency as

$$\eta_{EE} = \frac{\lim_{T \to \infty} \frac{1}{T} \sum_{t=0}^{T-1} \mathbb{E}\{E^{tol}(t)\}}{\lim_{T \to \infty} \frac{1}{T} \sum_{t=0}^{T-1} \sum_{i \in \mathcal{U}} \sum_{j \in \mathcal{B}} \mathbb{E}\{D_i^t(t) + D_{ij}^e(t)\}}. \tag{6.18}$$

This is the ratio of long-term total energy consumption to the corresponding long-term aggregate of accomplished computation tasks.

We define $\mathbf{a}(t) = [\mathbf{w}(t), \mathbf{p}(t), \Psi(t), \mathbf{f}^l(t), \mathbf{f}^e(t)]$ as the system action in time slot t, where $\mathbf{w}(t)$ is the bandwidth allocation vector, $\mathbf{p}(t)$ is the transmission power vector, $\Psi(t)$ is the vector associated with the computation task leaving the edge servers, and $\mathbf{f}^l(t)$ and $\mathbf{f}^e(t)$ are the vectors of computation resources that edge servers allocate to the devices. Taking the network stability constraint into account, the stochastic offloading problem for minimizing η_{EE} can be formulated as

$$\text{P1}: \quad \min_{\mathbf{a}(t)} \ \eta_{EE}$$

$$\text{s.t.} \sum_{i \in \mathcal{U}} \frac{w_{ij}(t)}{w_j} \leq 1, \quad w_{ij}(t) \geq 0, \tag{6.19a}$$

$$0 \leq p_i(t) \leq p_{i,max}(t), \tag{6.19b}$$

$$0 \leq f_i^l(t) \leq f_i^l, \tag{6.19c}$$

$$\sum_{i \in \mathcal{U}} f_{ij}^e(t) \leq f_j^e, \quad f_{ij}^e(t) \geq 0, \tag{6.19d}$$

$$\Psi_j(t) * c \leq f_j^e \tau, \quad \Psi_j(t) \geq 0, \tag{6.19e}$$

$$(6.12a) - (6.12b).$$

Constraint (6.19a) is the bandwidth allocation constraint. Constraints (6.19b) and (6.19c) denote the transmission power and computation resource constraints, respectively. Constraint (6.19d) is the computation resource allocation constraint. Constraint (6.19e) implies that the amount of computation resource for processing task Ψ_j cannot exceed the available computation resources.

Problem P1 is a stochastic optimization problem. The complex coupling among optimization variables and mixed combinatorials make P1 difficult to solve. Further, the stochastic task arrival, dynamic channel state information, and dynamic task buffer make it challenging to design an efficient resource management policy for the devices and edge servers. We therefore exploit Lyapunov optimization to transform the original stochastic optimization problem into a deterministic per-time block problem and propose a stochastic computation offloading algorithm to solve P1.

6.4.3 Lyapunov Optimization for Stochastic Computation Offloading

We define the quadratic Lyapunov function as the sum of the squared queue backlogs,

$$L(\Theta(t)) = \frac{1}{2}\{\sum_{i\in\mathcal{U}}[Q_i^l(t) - \beta_i]^2 + \sum_{j\in\mathcal{B}}Q_j^e(t)^2\}, \tag{6.20}$$

where $\Theta(t) = [Q^l(t), Q^e(t)]$ represents the current task queue lengths of the devices and edge servers, and β is a perturbation vector. Further, we define the Lyapunov drift-plus-penalty function as

$$\triangle_V L(\Theta(t)) = \triangle L(\Theta(t)) + V\mathbb{E}[\eta_{EE}(t)|\Theta(t)], \tag{6.21}$$

where $\triangle L(\Theta(t)) = \mathbb{E}[L(\Theta(t+1)) - L(\Theta(t))|\Theta(t)]$ is the conditional drift, and V is a non-negative weight parameter. By minimizing $\triangle_V L(\Theta(t))$, we can ensure network stability and simultaneously minimize network efficiency. The upper bound of $\triangle_V L(\Theta(t))$ can be derived as

$$
\begin{aligned}
\triangle_V L(\Theta(t)) \leq\ & C - \sum_{i\in\mathcal{U}}[Q_i^l(t) - \beta_i]\mathbb{E}[\Psi_i(t) - \lambda_i(t)|\Theta(t)] \\
& - \sum_{j\in\mathcal{B}}Q_j^e(t)\mathbb{E}[\Psi_j(t) - \sum_{i\in\mathcal{U}}D_{ij}^e(t)|\Theta(t)]\} + V\mathbb{E}[\eta_{EE}(t)|\Theta(t)],
\end{aligned}
\tag{6.22}
$$

where $C = \frac{1}{2}\{\sum_{i\in\mathcal{U}}[\Psi_{i,max}^2 + \lambda_{i,max}^2] + \sum_{j\in\mathcal{B}}[\Psi_{j,max}^2 + (\sum_{i\in\mathcal{U}}D_{ij,max}^e)^2]\}$, and $\Psi_{i,max}, \lambda_{i,max}, \Psi_{j,max}$, and $D_{ij,max}^e$ are the upper bounds of $\Psi_i(t), \lambda_i(t), \Psi_j(t)$, and $D_{ij}^e(t)$, respectively. Based on Lyapunov optimization theory, we can minimize the right side of the inequality in (6.22) to obtain the optimal solution of P1. Specifically, instead of solving P1, we can observe $\Theta(t)$ and $\lambda_i(t)$ to determine $a(t)$ by solving the following problem in each time slot:

$$
\begin{aligned}
\text{P2}: \min_{\mathbf{a}(t)}\ & V[E^{tol}(t) - \eta_{EE}(t)\sum_{i\in\mathcal{U}}\sum_{j\in\mathcal{B}}(D_i^t(t) + D_{ij}^e(t))] + \sum_{j\in\mathcal{B}} \\
& \{Q_j^e(t)[\sum_{i\in\mathcal{U}}D_{ij}^e(t) - \Psi_j(t)]\} - \sum_{i\in\mathcal{U}}[Q_i^l(t) - \beta_i][\Psi_i(t) - \lambda_i(t)] \\
& \text{s.t.}\ \ (6.12a) - (6.12b), (6.19a) - (6.19e).
\end{aligned}
\tag{6.23}
$$

Problem P2 needs to minimize the system cost per time slot. Here, we use DRL to solve P2, because it is efficient for finding a near-optimal solution in real time.

To solve P2, the system first constructs a Markov decision process, that is, $\mathcal{M} = (\mathcal{S}, \mathcal{A}, \mathcal{P}, \mathcal{R})$, and then uses a DRL algorithm to explore the actions. From Fig. 6.5, the network state $s(t)$ is constructed by digital twin and output to the DRL agent. To gather network information, digital twin needs to predict the locations, energy, and the generated task flow of the devices and BSs. The locations can be predicted by the K-nearest neighbours classification method in [76]. To prolong the battery

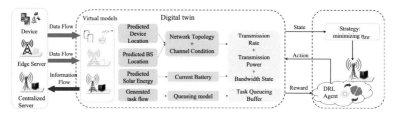

Fig. 6.5 Digital twin–enabled DRL

life of the devices, some of them are equipped with energy-harvesting chips, such as solar panels. Digital twin thus needs to support solar energy prediction here. The generated task flow is based on the application running on each device. Digital twins are used to first predict and gather the information on location, energy, and task flow. Then, based on the gathered information, digital twin updates the network topology, channel condition, and task queueing models. Finally, digital twin generates the current state and transmits it to the DRL agent.

The DRL agent constructs the system state as $s(t) = \{\mathbf{R}(t), \mathbf{F}, \mathbf{p}_{max}(t), \mathbf{w}, \Theta(t)\}$ with wireless data rate, computation resource, transmission power, and task queueing information. Action $a(t) = [\mathbf{w}(t), \mathbf{p}(t), \Psi(t), \mathbf{f}^{l}(t), \mathbf{f}^{e}(t)]$ is constructed with the bandwidth allocation, the transmission power, the executed computation task, and the computation resource allocation. It is worth noting that all the variables in action $\mathbf{a}(t)$ are continuous. Thus, we will utilize a policy gradient–based DRL algorithm to explore policy. After executing action $\mathbf{a}(t)$, digital twin updates the system state and estimates the immediate reward $\mathcal{R}^{imm}(s(t), a(t))$. Because the distribution of transition probabilities is often unknown in DRL, the DRL agent utilizes a deep neural network to approximate it. We define the immediate reward function $R^{imm}(s(t), a(t))$ as the objective of P2 problem. After computing the immediate reward, the system updates its state from $s(t)$ to $s(t + 1)$ based on action $a(t)$.

We use an online and asynchronous DRL algorithm to explore policy. The online DRL consists of a global agent and multiple learning agents. The detailed policy is explored by the learning agent in each SBS. The policy learned by the learning agent is $a(t) = \pi(s(t)|\theta_{\pi})$, where $\pi(s(t)|\theta_{\pi})$ is the explored offloading and resource allocation policy produced by a deep neural network. According to $s(t)$ and $a(t)$, the DRL agent can produce the reward and the next state. To estimate the performance of the proposed DRL algorithm, we consider a network topology with one MBS, $M = 3$ SBSs, and $N = 20$ devices. Each learning agent has an actor network and a critic network. The actor network has three fully connected hidden layers, each with 128 neurons, and an output layer with eight neurons using the softmax function as the activation function. The critic network has three fully connected hidden layers, each with 128 neurons, and one linear neuron output layer.

Figure 6.6 depicts the system costs with respect to training episodes under different schemes. The green curve is the benchmark of the joint optimization of computation offloading, the bandwidth, and the transmission power, but without computation resource allocation. The orange curve is the benchmark of the joint optimization of

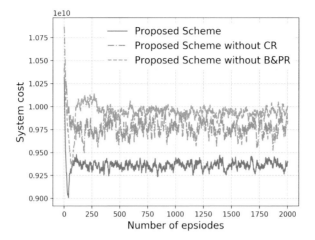

Fig. 6.6 System costs under different schemes

the computation offloading and computation resource allocation. Figure 6.6 shows that the performance of the proposed scheme outperforms the two benchmarks, since it can concurrently optimize computation offloading, the bandwidth, the transmission power, and the computation resource allocation. In addition, the system cost of the orange curve is lower than that of the green curve. This means that, compared with the optimization of computation resources, the joint optimization of the bandwidth and transmission power has a greater influence on performance.

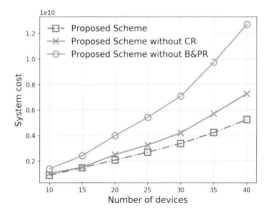

Fig. 6.7 System costs with respect to the number of devices under different schemes

Figure 6.7 compares the system costs with respect to the number of devices under different schemes. The number of devices ranges from 10 to 40. From Fig. 6.7, we can make two observations. First, for each of the three schemes, the system cost increases with the number of devices. The reason is that the increase of devices leads to more offloading requests, resulting in the consumption of more communication and computation resources. Second, the performance of the proposed algorithm outperforms two benchmarks by jointly optimizing computation offloading, bandwidth, transmission power, and computation resource allocation.

Chapter 7
Digital Twin for Aerial-Ground Networks

Abstract With the widespread deployment of unmanned aerial vehicles (UAVs) in civil and military fields, researchers have turned their attention towards the emerging area of aerial-ground networks for computing-intensive applications, data-intensive applications, and network-intensive applications. However, the application of aerial-ground networks relies on dynamic perceptions and intelligent decision making, which are difficult to conceive of due to the heterogeneity of ground devices and the complexity of the aerial-ground environment. The convergence of digital twin (DT) and UAVs has great potential to tackle the challenge and improve the service quality and stability in applications such as rescue and search and communication relaying. This chapter first investigates the advantages, challenges, and key techniques of DTs for aerial-ground networks. In addition, we highlight the main issues of DT for UAV-assisted aerial-ground networks with two case studies, including cross-domain resource management and intelligent cooperation among devices.

7.1 Introduction

Recently, aerial-ground networks based on unmanned aerial vehicles (UAVs) have made great success in various applications, such as disaster relief, service congestion, and damage assessment. Thanks to their inherent advantages, such as wide coverage, high flexibility, and strong resilience, UAVs can act as aerial mobile base stations to provide seamless and intelligent services for ground devices. However, due to the heterogeneity and mobility of ground devices and the dynamic network topology, the advantages of aerial–ground networks cannot be fully exploited. As an emerging digital mapping technology, digital twin (DT) has great potential to tackle the network dynamics and complexity of aerial-ground networks. By mapping the channel state and computing state, DT established on UAV can reflect the state of ground devices or network topology in a timely manner and accurately capture their state changes. After learning from these complex statuses, DT established on UAVs can support diversified applications, such as trajectory planning, large-scale mapping, urban

© The Author(s) 2024

Y. Zhang, *Digital Twin*, Simula SpringerBriefs on Computing 16,

https://doi.org/10.1007/978-3-031-51819-5_7

modelling, road patrol, and anti-piracy. We detail the main application scenarios of DT deployed on UAVs in different fields as follows.

- *Smart city*: In smart cities, with the help of DT deployed on UAVs, we can build a large-scale virtual city, dynamically monitor urban facilities, allocate urban public resources, and further realize intelligent collaborative decision making in urban management.
- *Disaster rescue*: In the disaster rescue field, UAVs with DT can analyse the connection performance of ground rescue devices and then make proactive communication resource allocations for high-priority devices to maximize the long-term quality of service (QoS).
- *Telemedicine*: In the field of telemedicine, through smart wearable devices, patients' health information can be sent back to DTs deployed on UAVs. UAVs with DT can track and monitor a patient's health status remotely and in a timely manner. When the DT measures any abnormal information, the rescue agency can immediately provide first aid services.
- *Internet of vehicles*: In the Internet of Vehicles (IoV), UAVs with DT can competently implement the real-time planning of vehicle trajectories in a specific area. At the same time, services such as status awareness and mobility prediction provided by DT can effectively avoid traffic congestion and reduce traffic accidents.

DT is able to assist in the optimal allocation and intelligent dispatching of valuable aerial resources. We further summarize the advantages of DT and UAV fusion as follows.

- *Hyperconnectivity*: Due to the wide coverage of UAVs over ground devices, DT deployed on UAVs can achieve interoperability and hyperconnectivity with physical counterpart devices. DT deployed on a UAV can connect all the ground devices in the aerial-ground network. We can fully utilize the advantages of DT from the multidimensional integration of information to sense how different devices work together, thus building an aerial-ground network with hyperconnectivity. In an aerial-ground network with hyperconnectivity, DT has the interaction details and status information of all the devices, which can then dynamically provide optimal decisions for different problems.
- *Low latency*: Thanks to the mobility of UAVs, DT deployed on a UAV can maintain a specified synchronization frequency with the ground device, which enhances the fidelity of the signal and brings more reliable DT services to the device. DT is sensitive to synchronization frequency, and untimely state synchronization or instruction updates can cause DT to make incorrect decisions. UAVs have the ability to move with mobile devices such as vehicles, significantly reducing DT status update delays due to communication distance. DT deployed on UAVs can better meet the requirements of devices for low network latency in different application scenarios, such as real-time trajectory planning in IoV, and provide services with higher performance and reliability.
- *Strong stability*: DT deployed on UAVs can monitor the status of each aerial-ground network in real time, which ensures the coverage and stability of the

aerial-ground network and makes the DT service more stable. Deploying DT on the ground makes it difficult to perform timely maintenance in the event of an attack or communication failure, which will lead to the interruption of DT services. DT deployed on a UAV can closely monitor the state changes in different aerial-ground networks, after detecting emergency situations such as UAV damage and network failure. DT can then immediately replenish and replace UAVs, continuously providing high-stability and high-performance services for devices.

The complementary advantages of DT and UAV play an important role in diverse applications that require stable network connections. However, there are still challenges in how to customize DT on UAVs for smart services in aerial-ground networks. We summarize the challenge in two cases.

- *Cross-domain resource allocation*: An aerial-ground network involves two different resource domains: the aerial domain and the terrestrial domain. The main challenge that DT faces on UAVs is the effective allocation of limited resources across domains under resource and distance constraints. DT-enabled intelligent services are often supported by a large amount of data distributed over various terminal devices. In a large-scale aerial-ground network, there are limitations of physical distance, communication resources, and computing resources; therefore, how DT deployed on UAVs effectively allocates resources across domains deserves in-depth study. In addition, the limited energy capacity of UAVs cannot support DT modelling, and DT relies on abundant computing resources and sufficient energy supply, which further limits the endurance of UAVs.
- *Cross-device intelligent collaboration*: The intelligent collaboration of different devices in an aerial-ground network is an important link to keep the network running efficiently. One of the important features of aerial-ground networks is a highly dynamic network environment. Diverse devices are constantly joining and withdrawing from the network, and mobile devices such as vehicles, UAVs, and mobile phones have low latency tolerance. For DTs deployed on UAVs, enabling different devices to achieve dynamic joint decision making and intelligent collaboration in tasks such as autonomous driving and trajectory planning while reducing network latency is challenging.

7.2 Key Techniques

7.2.1 Cross-Domain Resource Management

Aerial-ground networks can enhance the environmental perception and decision making capabilities of the network by leveraging multidimensional resources to achieve resource management. However, the resources in different domains (such as air and ground) are complicatedly coupled, and the orchestration of these cross-domain resources is confronted with a huge state–action space, which makes it difficult to allocate resources optimally in real time [78, 79]. To effectively manage the multidimensional resources (for communication, computing, and caching) of aerial-ground networks, the state change and QoS of the network are the key factors to consider.

Ensuring the flexibility and efficiency of resource management:

Aerial-ground networks are extremely dynamic and complex because of the high mobility of heterogeneous devices and the large scale of the networks. It is difficult to achieve flexible and efficient network resource management. As an emerging digital mapping technology, DT provides an approach for realizing effective and reliable network orchestration by mapping and predicting the dynamics of networks. Deng *et al.* in [80] proposed a combined approach of expert knowledge, reinforcement learning, and DT to cope with the dynamic changes of high-dimensional network states. Dai *et al.* in [74] proposed a new paradigm DT network for the Industrial Internet of Things (IIoT) and formulated random computing shunting and resource allocation problems, using Lyapunov optimization technology to transform the original problem into a deterministic per-slot problem. Lu *et al.* in [37] proposed a DT edge network to fill the gap between the physical edge network and the digital system. The integration of DT technology into aerial-ground networks can yield considerable improvement in both the latency performance and computing efficiency of applications running on ground devices and aerial devices.

Software-defined networking (SDN) can be utilized to construct and manage virtual networks to support specific network services for flexible network management [81]. Based on SDN architecture, Li *et al.* in [82] modelled multidimensional resource scheduling as a partially observable Markov decision process and used value iteration to jointly optimize networking, caching, and computing. Due to the complicated coupling of multidimensional resources, the central controller can hardly know a priori the effects of its actions on system performance. To this end, He *et al.* in [83] proposed a resource orchestration method based on deep reinforcement learning, with which the central controller learns an effective policy via trial-and-error search.

Ensuring QoS performance: The effective management of the multidimensional resources (for communication, computing, and caching) of aerial-ground networks to guarantee the required QoS performance of ground devices is also an important challenge. High computational complexity, the large cost of equipment deploymemt, and limited resources are the factors that hinder the improvement of QoS performance.

- *Reducing computational complexity*: Due to the limited computing and communication capabilities of ground devices, task offloading, as a key technology, can effectively improve service execution efficiency and realize the fast and efficient response of ground devices. Task offloading means that resource-constrained mobile terminal devices can offload overloaded computing tasks to edge nodes with stronger computing or communication capabilities, to improve computing speed and save energy. For example, road side units (RSUs) can undertake computation-intensive tasks (e.g. semantic image segmentation, motion planning, and route planning) for vehicles. Xu et al. in [30] proposed a service offloading method with deep reinforcement learning in DT-empowered IoV to provide vehicular services with a high QoS level. To reduce processing delays, Do-Duy et al. in [84] proposed a novel DT framework assisting in the task offloading of IoT devices for IIoT networks with mobile edge computing. Qu et al. in [85] proposed a deep meta-reinforcement learning offloading algorithm that combines multiple parallel deep neural networks with Q-learning, quickly and flexibly obtaining the optimal offloading strategy from a dynamic environment.

- *Reducing equipment deployment costs*: To achieve effective resource management and satisfy the diverse QoS requirements, the deployment cost of edge nodes cannot be ignored in aerial-ground networks. Using a large number of edge nodes to completely cover an area means a large deployment cost. When ground devices offload computing tasks to nearby edge nodes through the assistance of UAVs, the appropriate incentive is required for edge nodes to contribute their services. Edge nodes can be unwilling to contribute their services if the rewards cannot compensate for their service costs. Sun et al. in [86] designed an incentive mechanism to motivate RSUs to provide computing resources for ground vehicles. It was able to effectively complete vehicle task offloading schemes with the assistance of UAVs in an aerial-ground network. Zhou *et al.* in [87] proposed a novel incentive-driven and deep Q-network–based method and combined a content caching strategy and incentive mechanism to improve the performance of device-to-device offloading. To realize the long-term stability of DT services, Lin et al. in [88] designed an incentive-based congestion control scheme to offload real-time mobile data captured by DT to mobile edge computing servers.
- *Reducing the burden of aerial devices with limited resources*: Most works ignore the fact that centralized resource allocation schemes introduce a great burden to aerial devices, especially to UAVs in aerial-ground networks. Moreover, the incentive mechanism can be computation intensive, which results in service-unrelated energy consumption and further deteriorates service endurance. Thus, the resource allocation scheme should be carried out in a distributed manner. Through cooperative networks [89], SDN controllers can be decomposed into multiple simpler controllers to reduce the complexity of a large action space. Nasir *et al.* in [90] thus leveraged multi-agent deep Q-learning to distributedly schedule power allocation in wireless networks. The alternating direction method of multipliers (ADMM) is a distributed parallel optimization algorithm, and resource allocation problems based on ADMM have attracted much attention. Wang *et al.* in [91] considered computational offloading, resource allocation, and content caching strategies as optimization problems. An algorithm for solving optimization problems based on the ADMM algorithm was designed. Liang *et al.* in [92] proposed an efficient ADMM-based distributed virtual resource allocation algorithm in virtualized wireless networks. In addition, Zheng *et al.* in [93] designed a converged and scalable Stackelberg game–based ADMM for edge caching to solve storage allocation games and user allocation games in a distributed manner.

7.2.2 Cross-Device Intelligent Cooperation

In aerial-ground networks, heterogeneous ground devices can collaborate with aerial devices to accomplish intelligent network orchestration based on federated learning. Cross-device intelligent cooperation plays an important role in the efficient operation of networks and stable network environments. However, due to the heterogeneity, mobility, and selfishness of devices, across-device intelligent cooperation based on federated learning still faces many challenges. For example, further optimization is needed in terms of communication efficiency, training efficiency, and training costs. DT has the powerful ability to capture the state of heterogeneous devices in real time, which can effectively promote cross-device intelligent cooperation.

Improving communication efficiency: The heterogeneity and high mobility of devices complicate network management. The real-time changes of device states can lead to inaccurate channel estimation and affect the communication efficiency of federated learning. DT can analyse the connection performance of devices and make proactive communication resource allocations for improving communication efficiency. Lu *et al.* in [9] proposed a blockchain-based DT-enabled federated learning scheme to improve communication efficiency. Tran *et al.* in [94] studied the collaborative optimization problem when devices participate in federated learning in wireless networks. By adjusting a device's resource allocation strategy and the local training update frequency between two global aggregations, the best trade-off between communication time and computing performance can be achieved. Sun *et al.* in [33] used deep reinforcement learning to adaptively adjust the cooperative aggregation strategy of federated learning to achieve the balanced optimization of communication and computing. Krouka *et al.* in [95] proposed a novel distributed reinforcement learning algorithm to solve the random interference and communication interference of wireless channels and optimize communication efficiency.

Improving training efficiency: The dynamic nature of aerial-ground networks makes it difficult for heterogeneous devices to complete collaborative computing, so it is difficult to improve the training efficiency of federated learning. Lu *et al.* in [37] proposed a blockchain-empowered federated learning framework operating in a DT wireless network that comprehensively considers DT association, training data batchsize, and bandwidth allocation to formulate the training optimization problem. Jiang *et al.* in [36] exploited blockchain to propose a new DT edge network framework and designed a joint cooperative federated learning and local model update verification scheme that achieves the optimal unified time. Zhang *et al.* in [96] proposed a reinforcement of a federated learning scheme based on deep multi-agent reinforcement learning to optimize the training performance of federated learning in distributed IIoT networks. Li *et al.* in [97] proposed a platform-assisted collaborative learning framework. This framework can rapidly adapt to learning a new task at the target edge node by using a federated meta-learning approach with a few samples. Existing collaborative computing needs to restart learning as the topology changes, which leads to the failure or slow convergence of the established cooperative mechanism. DT can capture a complex network topology dynamically and improve the efficiency of collaborative computing between devices.

Reducing training costs:

It is necessary to encourage heterogeneous devices to participate in intelligent cooperation. Heterogeneous devices need to spend their resources and costs to train the federated learning model. They are therefore reluctant to participate in training without appropriate incentives [98, 99]. Existing incentive mechanisms can perform poorly due to the insufficient utilization of massive data and inaccurate modelling of operations in dynamic aerial-ground networks. DT can reduce the information asymmetry between devices by monitoring the status of devices in real time. Yang *et al.* in [100] introduced the Stackelberg game to establish an interaction model that comprehensively considers the data size, training time, and power consumption to measure the contribution to motivate client participation. Lim *et al.* in [101] studied the incentive mechanism for federated learning in UAV-assisted IoV to encourage contributions from data owners, considering information asymmetry between UAVs and the data owners. Federated learning is data driven, and the motivation of clients and the quality of data they provide have an important impact on the training results [102, 103]. The incentive mechanism combined with DT is suitable for motivating heterogeneous devices to actively participate in training in aerial-ground networks.

In summary, cross-device intelligent cooperation based on federated learning in an aerial-ground network still needs to be studied further. The integration of DT and aerial-ground networks can provide favourable support for realizing the cooperation mechanism of model-free, self-learning, autonomous intelligence.

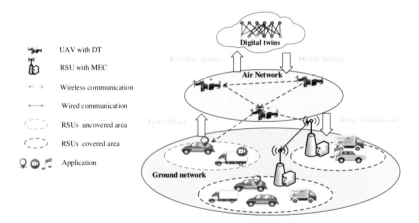

Fig. 7.1 A DT-driven aerial-ground network system model

7.3 DT for Task Offloading in Aerial-Ground Networks

7.3.1 System Model

To realize the efficient allocation of cross-domain resources from air and ground, we establish a dynamic DT model for an aerial-ground network. The DT model can capture the time-varying demand and supply of cross-domain resources in the network. Deploying the DT model on devices in the network can significantly improve the environmental perception, computing efficiency, and delay performance of the devices. This is beneficial for the unified and efficient resource allocation and scheduling in aerial-ground networks.

As shown in Fig. 7.1, we consider a DT-driven aerial-ground network in which vehicles act as ground devices and UAVs act as aerial devices. The network is composed of vehicles, RSUs, UAVs and DTs. We assume the UAVs are responsible for areas that are not covered by RSUs, as a supplement to the ground network. In such areas, vehicles and RSUs are able to deliver messages to a UAV directly with line-of-sight communication. With the assistance of UAVs, the vehicles not covered by the ground network could offload their computing tasks to RSUs to reduce their computing burden. We establish two DT models, including the DT of a group of RSUs and the DT of vehicles. Both DTs are established in UAVs to update the network topology and traffic load in real time and help UAVs make specific decisions, such as path planning. The DT of a group of RSUs can be given by

$$\mathcal{D}^r = \{\mathcal{F}^r, G^r, L^r\}, \tag{7.1}$$

where $\mathcal{F}^r$ is a vector describing the available computing resource status of the RSUs, G^r is the network topology between the RSUs, and L^r is the network transmission load of the RSUs.

The DT of vehicles can be given by

$$\mathcal{D}^v = \{G^v, L^v, C, Q\}, \tag{7.2}$$

where G^v is the network topology of the vehicles, L^v is the communication load of the vehicles, C represents the demand information of the vehicles at this time, and Q is the preference of the vehicles for the resource providers. The preference is determined by the historical service of the vehicles in a specific type of offloading task.

7.3.2 Utility Function

The DT of a group of vehicles and the DT of a group of RSUs have different utilities. The set of RSUs in the network is $\mathcal{M} = \{1, ..., m, ..., M\}$. The set of vehicles that require offloading tasks is $\mathcal{N} = \{1, ..., n, ..., N\}$. Vehicle n wants to maximize its service satisfaction, which is the accumulated satisfaction it achieves from various RSUs. A vehicle's satisfaction is defined as the ratio of its cumulative satisfaction from RSUs to the total number of resources it receives. Thus the satisfaction of vehicle n is given by

$$S_n = \frac{\sum\limits_{m \in M} \{q_{n,m} p_{m,n} - \frac{q_{n,m} p_{m,n}^2}{2\tilde{f}}\}}{\sum\limits_{m \in M} p_{m,n}}, \tag{7.3}$$

where $\tilde{f}$ is the maximum expected value of resources from RSUs, $p_{m,n}$ represents the CPU frequency obtained by vehicle n at RSU m, and $q_{n,m}$ represents the preference of vehicle n for RSU m.

The DT of RSUs tries to minimize energy consumption. The energy consumption on the RSU is related to the frequency and duration of the CPU used. We can express energy consumption as

$$E(\mathcal{P}) = \sum_{m \in M} \sum_{n \in N} \omega p_{m,n}^2 c_{n,m}, \tag{7.4}$$

where ω represents the effective capacitance parameter of the computing chipset, and $c_{n,m}$ is the number of CPU cycles required for RSU m to calculate its tasks for vehicle n. The detailed resource scheduling of each vehicle is expressed as $\mathcal{P} = \{P_m, m \in M\}^T$.

7.3.3 Distributed Incentives for Satisfaction and Energy Efficiency Maximization

The goals of RSUs and the DT of RSUs are different. An RSU is designed to maximize the average satisfaction of the vehicle. whereas an RSU's DT is designed to maximize global energy efficiency. Although these quantities are used to formulate the allocation scheme of computing resources, it is difficult to achieve the goal of

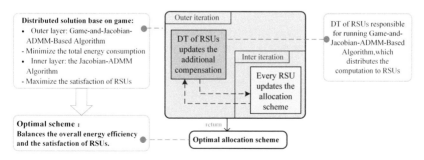

Fig. 7.2 Workflow of a game and Jacobian ADMM-based algorithm

minimizing total energy consumption when they have different optimal values. In addition, due to limited computing resources, computationally intensive centralized computing creates pressure for UAVs. Therefore, we propose an incentive mechanism based on the Stackelberg game and Jacobian ADMM to allocate computing resources, so that the DT of RSUs and the RSUs can reach a consensus on the allocation scheme and solve the whole problem in a distributed and parallel manner.

Due to the complexity of solving the desired objectives of RSUs and the DT of RSUs, we first derive the optimization problem of RSUs and the DT of RSUs and then construct a Stackelberg game. We solve the average satisfaction maximization problem for vehicles and the global energy efficiency maximization problems for the DT of RSUs by using the classic ADMM and Jacobian ADMM with two blocks, respectively. We obtain the resource allocation schemes of the two problems (the DT-driven classic ADMM and the DT-driven Jacobian ADMM). Furthermore, we model these two problems as a complete Stackelberg game. In the game, the RSUs' DT is the leader and the RSUs are the follower. According to the goals of RSUs and the DT of RSUs, we can formulate the Stackelberg game as

$$
\begin{aligned}
Leader: \quad & \underset{\mathcal{P}}{minimize} \quad E\left(\mathcal{P}\right) \\
Follower: \quad & \underset{p_m, \overline{Q}_m}{minimize} \quad \Phi_m(h_m(P_m, \eta_m), \boldsymbol{\theta}_m) \\
& \text{s.t.} \quad \sum_{n \in N} p_{m,n} = f_m, m \in \mathcal{M} \quad (C1),
\end{aligned} \tag{7.5}
$$

where $\overline{Q}_m$ is the cumulative preference of all the vehicles for RSU m. The term $\Phi_m(\cdot)$ includes the optimization direction of the DT of the RSUs and RSU m and the compensation from the DT of the RSUs. The classic ADMM with two blocks is powerless in this kind of convex optimization problem with high-dimensional variables. The Jacobian ADMM-based algorithm is able to solve convex optimization problems by breaking them into smaller subproblems, making each part more tractable. Therefore, we use the game and Jacobian ADMM-based algorithm to solve the problem. The algorithm flow is shown in Fig. 7.2.

In the beginning, the DT of the RSUs, as the leader, sends the incentive parameter θ_m to the corresponding RSU m, that is, the additional compensation of the DT of the RSUs to RSU m. We define the number of iterations of the outer loop as k. At iteration k, given incentive parameters $\{\boldsymbol{\theta}_1, \cdots, \boldsymbol{\theta}_m, \cdots, \boldsymbol{\theta}_M\}$ from the leader, each RSU updates its own computing resource allocation scheme P_m in the inner loop, and then the leader and the follower can reach the current optimal scheme. At the next iteration, $k + 1$, the leader will adjust the incentive parameters based on the updated $P_m, \forall m \in \mathcal{M}$. Then, a new current optimal scheme can be reached. When the outer iteration is terminated, the optimal incentive parameters and resource allocation scheme are the equilibrium point $(\theta^*, \mathcal{P}^*)$ of the Stackelberg game. The proposed DT-driven game ADMM minimizes global energy consumption based on the premise of ensuring the satisfaction of the RSUs.

7.3.4 Illustration of the Results

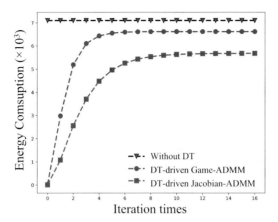

Fig. 7.3 Convergence of the energy consumption of all RSUs over iterations

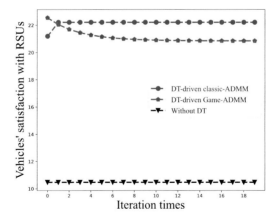

Fig. 7.4 Vehicle satisfaction with RSUs over iterations under three schemes

Figure 7.3 compares the energy consumption of three schemes, that is, the DT-driven Jacobian ADMM, the DT-driven game ADMM, and the scheme without DT, over the numbers of iterations. The scheme without DT allocates resources without the preference information that was obtained from the DT of the vehicles.

The energy consumption of the scheme without DT is the highest and remains a constant, because the tasks and CPU frequency can only be allocated randomly. This leads to a decision making and optimization process without iteration. The energy consumption of the DT-driven Jacobian ADMM is the lowest, since minimizing total energy consumption is its only objective at the cost of low vehicle satisfaction. The proposed DT-driven game ADMM jointly considers the overall energy efficiency and the satisfaction of the RSUs, and its energy consumption is thus higher than that of the DT-driven Jacobian ADMM.

Figure 7.4 compares the vehicles' satisfaction with the RSUs of three schemes, that is, the DT-driven classic ADMM, the DT-driven game ADMM, and the scheme without DT. Due to the contradictory goals of the RSUs and the DT of the RSUs, the DT-driven game ADMM attempts to balance between the two contradictory goals, and its satisfaction is a bit lower than that of classic ADMM. This is because RSUs allocate a great deal of resources to vehicles with high preferences, to provide satisfactory services for the vehicles. The satisfaction achieved by both DT-driven schemes, that is, the DT-driven Jacobian ADMM and the DT-driven game ADMM, is much higher than that without DT. This is because, in the scheme without DT, the preferences of the vehicles for RSUs are unknown, and the allocation cannot fully meet the actual requirements of the vehicles.

7.4 DT and Federated Learning for Aerial-Ground Networks

7.4.1 A DT Drone-Assisted Ground Network Model

Figure 7.5 shows a drone-assisted ground network scenario consisting of drones, ground clients, and DTs, where the drones provide supplementary capacity for ground communications during natural disasters or traffic peaks. Mobile drones with a wide range of coverage act as servers, responsible for task offloading, global model updates, and so forth. A wide variety of ground equipment, such as smartphones and laptops, serves as clients to perform tasks and connect with drones through wireless communications.

The drone serving as the aggregator cooperates with the ground equipment serving as the trainers to perform federated learning tasks. The drone publishes a global model ω, which all participating clients will download. Then, each client uses its own private data sets to train the model and upload the new weights or gradients to the server. This process is conducted iteratively until the entire training process converges [104, 74].

The establishment of DT can capture the state of network elements in real time and effectively help the system make intelligent decisions. DT types include the DT of ground clients and the DT of the drone. The DTs of ground clients are deployed on a resource-rich ground node. The drone would maintain the DT by exchanging information with the ground node instead of all the clients. The set of clients in the network is $\mathcal{N} = \{1, 2, \cdots, N\}$. Client i's DT, DT_i^c, at time t can be expressed as

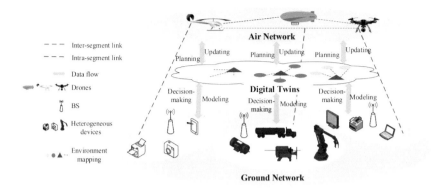

Fig. 7.5 The architecture of a DT-empowered aerial-ground network

$$DT_i^c(t) = \{F_i^t(\omega), b_i(t), f_i(t)\}, \tag{7.6}$$

where ω denotes the current training parameter of client i, $F_i^t(\omega)$ represents the current training state of client i, $b_i(t)$ represents the packet loss rate, and $f_i(t)$ is the CPU frequency of the client at time t.

Due to the deviation of DTs, the packet loss rate deviation $\hat{b}_i(t)$ and the CPU frequency deviation $\hat{f}_i(t)$ can be measured as the errors of the DT mapping in the communication environment and computing power, respectively. For client i, the calibrated DT is

$$\hat{DT}_i^c(t) = \{F_i^t(\omega), b_i(t) + \hat{b}_i(t), f_i(t) + \hat{f}_i(t)\}. \tag{7.7}$$

The DT of the drone manages the deviation of the DTs of the clients and has a preference for the clients. Drone j's model is

$$DT_j^u(t) = \{\mathcal{P}(t), \hat{\mathcal{D}}(t)\}, \tag{7.8}$$

where $\mathcal{P}(t)$ is the reputation distribution of nodes within its coverage area, and $\hat{\mathcal{D}}(t)$ is the set of deviations between the client's local update and the global update.

7.4.2 Contribution Measurement and Reputation Value Model

Update significance can intuitively measure the contribution of a local model update to the global model update. The update significance is measured by the model deviation d_i^τ, which is the divergence of a particular local model from the average across all local models. A small d_i^τ reflects a high quality of upload parameters of client i. The aggregator updates the value of d_i^τ for client i in each time slot, as a basis for the quality evaluation of the parameters submitted by client i.

The reputation of a client can also affect the training process. Through the reputation model, high-performance clients should be identified in terms of sufficient communication resources, powerful computing capabilities, and accurate training results. We use $\mathcal{P} = (\rho_1, \rho_2, \cdots, \rho_N)$ to represent the reputation value of each client. According to subjective logic, the reputation value model is related to the communication capability of node i during the τth global update and the learning quality d_i^τ.

7.4.3 Incentive for Federated Learning Utility Maximization

Static and dynamic incentives are designed for small-scale networks and large-scale networks, respectively [37]. In a small-scale network, a single drone can cover all the clients. Therefore, we first design a static incentive mechanism. The term τ_i represents the decision of client i, that is, the number of rounds in which the client participates in the global update; $\mathcal{T} = (\tau_1, \tau_2, \cdots, \tau_N)$ represents the strategies for all the clients; and $\tau_{-i} = (\tau_1, \cdots, \tau_{i-1}, \tau_{i+1}, \cdots, \tau_N)$ denotes the training strategies of all the clients except for client i. Given the computing cost per round (a complete global update round) $C = (c_1, c_2, \cdots, c_N)$ and the communication cost per round $\mathcal{K} = (k_1, k_2, \cdots, k_N)$, the static incentive utility function is the difference between the reward and loss of client i, which can be defined by

$$U_i(\tau_i, \tau_{-i}) = \frac{\rho_i \tau_i}{\sum_{j \in N} \rho_j \tau_j} R - \tau_i c_i - \tau_i k_i. \tag{7.9}$$

The utility function of the aggregator is the total energy consumption of clients in the learning process minus the payment of the aggregator. The static incentive utility function is defined as

$$U_0(R) = \sum_{i \in N} \rho_i \tau_i c_i - \alpha R^2, \tag{7.10}$$

where $\alpha > 0$ is a system parameter to ensure that the utility is greater than or equal to zero under the optimal R^*.

In a large-scale case, it is difficult for a single drone to cover the entire area. Therefore, a dynamic incentive mechanism can be designed to select the optimal clients in adaptation to the time-varying environment. The difference from the static incentive is that C in the dynamic incentive represents the computing cost of the client to complete a round of local training. In the dynamic scene, we use r^τ instead of R in the formula, where r^τ represents the reward determined by the drone before the τth global model is updated. For convenience, in the following analysis, we uniformly use R to express the reward.

The decision making problem can be modelled using the Stackelberg game. In the game, the DT of the drone acts as the leader, while the ground clients are the follower. The game consists of two stages. In the first stage, the aggregator publishes the task and determines its reward R. In the second stage, each client will devise strategies to determine the number of rounds to participate in federated learning and maximize their respective utilities [105]. The second stage of the Stackelberg game is a noncooperative game, that is, in which there is a Nash equilibrium. A set of strategies $\mathcal{T}^* = (\tau_1^*, \tau_2^*, \cdots, \tau_N^*)$ is a Nash equilibrium in the second stage of the game if, for any client i, $U_i(\tau_i^*, \tau_{-i}^*) \geq U_i(\tau_i, \tau_{-i}^*)$, $\forall \tau_i > 0$. Under the reward R given by the aggregator, no client can gain any additional benefits by unilaterally changing the current strategy.

According to Nash equilibrium, when all the other clients expect client i to play their best strategy, client i can only play τ_i^*. Therefore, we need to introduce the concept of the best response strategy. Given τ_{-i}, a strategy is client i's best response strategy, denoted by $\beta_i(\tau_{-i})$, if it maximizes $U_i(\tau_i, \tau_{-i})$ over all $\tau_i \geq 0$. To find the Nash equilibrium in the second stage of the game, a closed-form solution of the best response strategy for each client must be calculated. Accordingly, if the whole game has a unique Stackelberg equilibrium, the necessary and sufficient condition is for there to be a unique optimal solution in the first stage of the game. There exists a unique Stackelberg equilibrium $(R^*, \mathcal{T}^*)$, where R^* is the only value that can maximize the utility of the aggregator over $R \in [0, \infty)$. The utility function of the aggregator is a concave quadratic function on the difference between the reward and loss of client i, and the first derivative of the utility function is equal to zero. Then the optimal R can be solved. At this time, $(R^*, \mathcal{T}^*)$ is the unique Stackelberg equilibrium in the game.

Different from the static mechanism, the dynamic mechanism selects clients according to the ratio of the unit local training computing cost and reputation value, that is, $\frac{c_i}{\rho_i}$. The drone's optimal payment R^* should be expressed as $R^* = \sum_{\tau=1}^{\tau_g} (r^\tau)^*$, where τ_g is the number of rounds of the global update. Finally, t_i^* and $(r^\tau)^*$ constitute the unique equilibrium of the Stackelberg game.

7.4.4 Illustration of the Results

We use the software Pytorch 0.4.1 to build a federated learning model in an air–ground network and use the classic Modified National Institute of Standards and Technology data set to evaluate the performance of the proposed incentive mechanisms. We set up a total of 10 to 100 clients. Under the dynamic incentive, the communications range of a drone can only cover 20 clients at the same time. We employ a cost-only scheme as the benchmark where clients with low training costs are selected to participate in federated learning.

Figure 7.6 shows the model's accuracy with varying global update rounds under three schemes. The convergence accuracy of the global model relies on the participating clients and their data quality. The accuracy under the dynamic incentive scheme is the highest. After each round of global updates, the performance of the clients will be evaluated, and the participation of low-quality clients will be reduced.

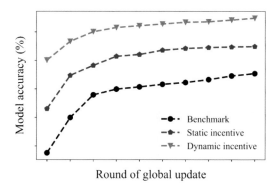

Fig. 7.6 Comparison of model accuracy under varying global update rounds

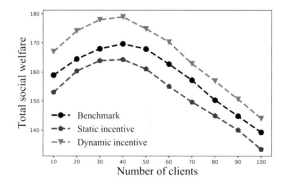

Fig. 7.7 The total social welfare of a drone and clients varies with the number of clients

The static scheme chooses the optimal client set, which might not be appropriate later in the federated learning process due to the mobility of the drone. Thus, the accuracy of the static incentive is lower than that of the dynamic incentive. Since the benchmark considers only the training costs of the clients, its model accuracy is 5% lower than that of the static scheme.

Figure 7.7 compares the total social welfare of the drone and clients varies with the number of clients under three schemes. As the total number of clients increases, the total social welfare increases first, peaks at around 40 clients, and then decreases. With the increase of the client number, the utility of the drone increases, while the utilities of the clients decrease due to the greater number of competitors. In addition, the benchmark social welfare is higher than that of the static incentive, because the benchmark selects only clients with low cost. Thus, its social welfare is the highest among the three schemes.

Chapter 8
Digital Twin for the Internet of Vehicles

Abstract As a working combination of smart vehicles, advanced communication infrastructures, and intelligent transportation units, the Internet of Vehicles (IoV) has emerged as a new paradigm for safe and efficient urban life in the future. However, various types of smart vehicles with distinct capacities, diverse IoV applications with different resource demands, and unpredictive vehicular topology pose significant challenges to fully realize IoV systems. To cope with these challenges, we leverage digital twin (DT) technology to model complex physical IoV systems in virtual space, to identify the relation between application characteristics and IoV services, which facilitates effective service scheduling and resource management. In this chapter, we discuss the motivation, benefits, and key issues of applying DT in IoV systems. Then, we use vehicular edge computing and caching as two typical IoV application scenarios to present DT-empowered task offloading and content caching scheduling schemes and their performance.

8.1 Introduction

Vehicles are undergoing a fundamental shift, from simple transportation units to smart ones empowered with environmental sensing, autonomous driving, and information interaction capabilities. Integrating such smart vehicles with pedestrians and the infrastructures around them gives rise to Internet of Vehicles (IoV) systems, which provide a range of powerful vehicular applications and lead to pioneering advances in safety and the efficiency of intelligent transportation. For instance, IoV helps to deliver information gathered from the urban traffic environment to adjacent vehicles for safe navigation and traffic management. In addition, IoV can provide real-time information and interactive entertainment for vehicle occupants.

The development of IoV technology has received much attention in recent years, and high expectations have been raised about the benefits that its application will bring, prompting researchers and engineers to engage in in-depth discussions on possible obstacles in the IoV evolutionary graph.

Y. Zhang, *Digital Twin*, Simula SpringerBriefs on Computing 16,
https://doi.org/10.1007/978-3-031-51819-5_8

The key feature of IoV is its massive connections and dynamic topology. As we mentioned, IoV is a network consisting of vehicles, drivers, pedestrians, roadside units (RSUs), and other intelligent units participating in traffic applications, communicated in vehicle-to-vehicle (V2V), vehicle-to-RSU (V2R), vehicle-to-person, and vehicle-to-sensor modes. The mobility of vehicles and pedestrians can cause drastic changes in data transmission performance and even the interruption of communication links. The large scales of connected units and time-varying communication associations make IoV characteristic modelling and operation management seriously complex and difficult.

Another issue worth considering is the ultra-low latency constraints of some IoV applications. For example, in vehicle driving, when the vehicle in front brakes in an emergency, the following autonomous vehicle needs to complete the braking action within a few milliseconds according to the detected vehicle distance or the warning notification sent by the front vehicle. To meet such a strict delay constraint, the vehicle's control, environmental perception, vehicular communication, and information processing must be comprehensively coordinated.

The last issue to be addressed is closely related to the previous one. Different types of IoV applications rely on different forms of cooperative services from heterogeneous resources. For example, vehicular augmented reality needs to consume a great deal of computing and sensing resources, while onboard interactive entertainment mainly relies on communication and storage resources. Furthermore, synergy and competition exist between heterogeneous resource services. For instance, the premise of data processing is that the data can be transmitted to the corresponding processor node by communication resources, which may be in contention due to multiple vehicle communication pairs. The complex relation between these resources makes it challenging to efficiently implement IoV applications.

Several technical approaches to the above challenges have emerged, with Digital twin (DT), in particular, showing promise. By mapping physical IoV networks to virtual space, DT helps improve IoV application performance and resource efficiency. Some of the main benefits provided by DT to IoV are shown in Fig. 8.1 and are listed below.

Accurate mapping and unified modelling: In DT-empowered IoV networks, DT servers collect road traffic status and application service characteristics from sensors installed on smart vehicles and through communication facilities spread throughout the vehicular network, to construct a real-time and accurate reflection of physical IoV networks. Since a reflection model in virtual space is represented by multidimensional digital parameters, irrelevant physical difference between various types of vehicles can be shielded by normalizing the feature parameters, to build a unified model that enables modelling interaction and migration.

Feature digging and trend prediction: In the process of autonomous driving and on-board application services, vehicles can consume various resources, such as urban roads, vehicular communications, and edge computing. Therefore, collaboration, competition, and even social associations among multiple vehicles are generated. DT reflection helps to explore such potential features and relations in IoV systems.

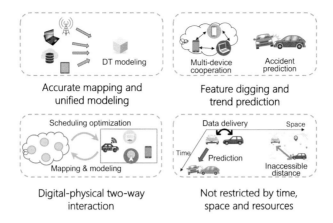

Fig. 8.1 Benefits provided by DT to IoV

Going a step further, based on these relations, DT can predict future physical actions, states, and events in IoV systems, such as possible traffic congestion or collisions.

Digital–physical two-way interaction: There is a two-way interaction between the DT model and the real physical entities of the IoV system. On the one hand, physical entities determine the digital mirroring. On the other hand, digital models logically guide physical action strategies. Both model accuracy and physical strategy performance can be improved during this iterative evolution process.

Not restricted by time, space, or resources: In physical IoV networks, the safety predictions of vehicle driving behaviour, inter-vehicle communications, and resource cooperation between vehicles are restricted by event sequences, wireless transmission distances, and vehicle resource capacities, respectively. However, in the DT image of IoV, these constraints can be broken. For example, by dynamically changing the timeline, retrospective determinations and predictions of traffic events are convenient to make. In addition, in virtual space, communications between inaccessible vehicles can be realized by data sharing between vehicle model processes in a DT server.

Motivated by the potential benefits of DT technology, a few works have addressed the incorporation of DT into IoV systems. In [106], the authors leveraged DT to facilitate collaborative and distributed autonomous driving. Based on vehicle DT models, driving decisions can be obtained at low cost. In [107], two DT models of vehicle driving states based on a Gaussian process and deep convolutional neural networks were respectively established that provide a scheme for the optimization of vehicle driving states and the realization of DT entity interactions. The authors in [108] introduced a DT-enabled edge intelligent cooperation scheme that guides optimal edge resource allocation and edge intelligent cooperation. Combining DT with vehicle-to-cloud communications, the authors in [109] presented a cooperative ramp merging system for connected vehicles that allows merging vehicles to cooperate with others prior to arriving at a merging zone. In [110], the authors focused on the security issues of cooperative intelligent transportation systems and constructed a DT model based on convolutional neural networks and support vector regression. Aided by the DT model, system security prediction accuracy was improved.

Despite much promising recent work in the area of DT-empowered IoV, several questions remain open for further investigation, and are discussed below.

Delays in DT modelling: Traffic safety is an important application scenario of DT-enabled IoV, in which some functions, such as early warnings of upcoming traffic accidents and adjustments of vehicle driving behaviour, have strict delay constraints. Meeting these constraints requires the DT model for the traffic environment and vehicle state to be constructed in a short time and to remain updated in real time. Considering the highly dynamic IoV topology and massive amounts of connected IoV nodes, the maintenance and tracking of such a complex system in real time are a challenge.

Efficiency in DT modelling: Following the previous challenge, to reduce DT modelling delays, many resources need to be allocated for vehicular environment sensing, state information delivery, and modelling processing. However, in addition to serving in the construction and update of DT models, constrained IoV resources are also used to support vehicular communication, autonomous driving, and onboard multimedia applications. How to reduce the resource costs of DT modelling and improve DT efficiency has become an important issue to be investigated.

Fault tolerance in DT modelling: The last but not least question concerns fault tolerance in DT modelling. Due to a limited sensing range, vulnerable wireless transmission parameters, and poor modelling processing power, established IoV DT models can have errors. These errors can seriously affect the control of vehicles' driving action and mislead the prediction of road traffic trends, thereby undermining the safety and efficiency of road traffic. In a harsh IoV environment, how to construct a DT model with high fault tolerance is still an unexplored problem.

8.2 DT for Vehicular Edge Computing

Driven by advances in vehicular communication and sensing and processing capabilities, many powerful IoV applications have emerged, such as autonomous driving, smart logistics, and driving augmented reality. However, the implementation of these applications requires intensive computation for environmental information processing and obtaining traffic behaviour under strict delay constraints, posing great challenges for vehicles with limited onboard computing resources.

VEC, which enables computing resource sharing at the edge of vehicular networks, is an appealing paradigm for meeting the intensive computation demands. In VEC, resource-hungry vehicles can offload their computing tasks to other smart vehicles or an RSU with spare computing power. However, to achieve efficient task offloading in such a dynamic and complex IoV environment, key issues still need to be addressed. For example, the communication scheduling for task data delivery is closely related to the computing resource management for task processing, which makes task offloading complicated. Moreover, resource competition between different offloading vehicle pairs, as well as the time-varying topology of vehicular networks, introduces further unprecedented challenges in managing VEC.

Recent advancements in machine learning provide significant capabilities to an aware dynamic IoV environment, determine action strategies, and tackle complex problems that rely in VEC applications. However, the effective implementation of the learning approach always relies on accurate and real-time system information gathered by learning agents. In vehicular networks characterized by massive amounts of connected smart vehicles, a highly dynamic topology, and a limited wireless spectrum, it is impractical to form a centralized artificial intelligence (AI) manager that schedules edge services for the entire network. To address this problem, we turn to multi-agent distributed learning empowered vehicular edge management. However, efficient collaboration and joint decision optimization among these multiple agents still face critical challenges.

DT is a promising technology to address these challenges. DT's state mapping between real and virtual dimensions provides users with comprehensive insights into the investigated system and dramatically reshapes the design and engineering process. Merging DT with machine learning will generate great benefits. On the one hand, DT provides AI with comprehensive and accurate system state information, which is exactly what learning processes require. On the other hand, AI provides much intelligence to DT, making its information collection and system description smart and efficient.

In this section, we propose a new VEC network based on DT and multi-agent learning that improves agent collaboration and optimizes task offloading efficiency [72]. In this network, DT is leveraged to reveal the potential cooperation between different vehicles and adaptively form multi-agent learning groups, which reduces learning complexity. Moreover, we design a distributed multi-agent learning scheme that minimizes vehicular task offloading costs under strict delay constraints in complex vehicular networks and dynamically adjusts the state-mapping mode of the DT network (DTN).

8.2.1 System Model

Figure 8.2 shows the framework of a DT-empowered VEC network. There are N smart vehicles on the road. These vehicles are equipped with computing power to process tasks and perform learning functions. The computing capability of vehicle i, $i \in \mathcal{N}$, is denoted as f_i CPU cycles per second. To enable powerful vehicular applications, such as autonomous driving and onboard entertainment, vehicles generate various types of tasks to be processed. Without loss of generality, we consider vehicle i to have J_i types of tasks, and task $w_{i,j}$ is described in the form of three elements, as $w_{i,j} = \{C_{i,j}, D_{i,j}, T_{i,j}^{\max}\}$. Here, $C_{i,j}$ is the amount of computing resources required to execute the task, $D_{i,j}$ presents the size of the task input data, and $T_{i,j}^{\max}$ is the maximum delay that task $w_{i,j}$ can tolerate.

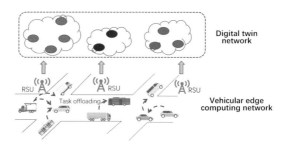

Fig. 8.2 A DT-empowered VEC network

Since different vehicles have diverse computing capabilities and task processing requirements, parts of the vehicles can have sufficient computing resources, whereas others are lacking. Through V2V communication, one vehicle can offload its tasks to others. We call the target vehicles "vehicular edge servers". Let $\beta_{i,j,k} = 1$ denote vehicle i, which offloads its task j to vehicular server k, and $\beta_{i,j,k} = 0$ denotes when the vehicle does not offload task j to server k. The time consumed to complete task $w_{i,j}$ is divided into two parts, namely, the offloading task transmission time and the task execution time. The transmission time of task $w_{i,j}$ from vehicles i to k through channel l is shown as $T_{i,j,k,l}^{\text{tran}} = D_{i,j}/R_{i,k,l}$, where $R_{i,k,l}$ is the transmission rate.

A target vehicular server can receive multiple tasks from the other vehicles, and it puts these tasks in a queue. Taking into account task delay constraints, the target server executes the tasks in order according to the length of remaining time, from shortest to longest. Consequently, a task's execution time consists of the waiting time in the queue and the time processed in the CPU. The execution time of $w_{i,j}$ can be presented as

$$T_{i,j,k}^{\text{exe}} = \sum_{i'=1}^{N} \sum_{j'=1}^{J_{i'}} \mathbf{1}\{T_{i',j'}^{\text{rem}} \le T_{i,j}^{\text{rem}}\}\beta_{i',j',k} C_{i',j'}/f_{i'}, \tag{8.1}$$

where $\mathbf{1}\{\hat{x}\}$ is an indicator function that equals one if $\hat{x}$ is true, and zero otherwise, and $T_{i,j}^{\text{rem}}$ is the remaining time of task $w_{i,j}$ before the deadline.

To improve vehicular computing resource utilization, a price-based incentive mechanism is incorporated into the resource scheduling. For a vehicular server, the weaker its computing power, the greater the resource demands of its queuing tasks, the tighter the tasks' delay constraints, and the higher the price of resources providing for guest tasks. We denote the price of a unit of computing resource of vehicle i as z_i.

In the vehicular edge system, a DTN continually maps the vehicles' physical states, such as the communication topology and computing resource demands, to virtual digital space. With the help of the DTN, edge service optimization and resource allocation strategies can be efficiently obtained.

8.2.2 DT and Multi-Agent Deep Reinforcement Learning for VEC

Merged with DT, AI learning gains comprehensive state information and effective guidance for agent learning, while helping DT to accurately model the physical system. We investigate the incorporation of DT and multi-agent learning in VEC networks and propose optimal edge service scheduling schemes. The main framework of these schemes is shown in Fig. 8.3.

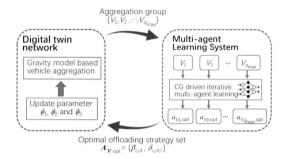

Fig. 8.3 Incorporation of DTN and multi-agent learning for VEC

Owing to the large-scale distribution of massive numbers of vehicles, it is costly and impractical to globally schedule the task offloading of the whole edge network. To address this issue, we leverage a DTN and gravity model to design an edge service aggregation scheme that efficiently aggregates vehicles based on the potential matching relations between the supply and demand of computing resources and greatly reduces the complexity of task offloading scheduling.

To guide the edge service aggregation, DTNs of the vehicular edge network are constructed in the RSUs. A DTN can be regarded as a combination of logical models and parameters recorded in digital space to characterize the states of the objects in physical space. We define the element of a DTN as $D_s = \{\mathcal{M}, \Phi, \varpi\}$. Here, $\mathcal{M}$ denotes the digital model of the vehicles in the physical system, which is described by a vehicle task set $\{w_{i,j}\}$, a computing capability set $\{f_i\}$, a resource price set $\{z_i\}$, and an available transmission rate set $\{R_{i,j}\}$. The modelling parameters are $\Phi = \{\phi_1, \phi_2, \phi_3\}$, which reflect the importance of the three factors of resources, pricing, and communication in the DTN modelling, respectively. The values of the parameters update periodically, and ϖ is the sequence number of the mapping periods.

With the aid of the DTN, we develop a gravity model–based vehicle aggregation scheme. Here we reform the gravity model and make it suitable to characterize the supply and demand relations of the vehicular edge service. The gravitation in the service association between vehicles i and i' is calculated as

$$F_{i,i'} = \frac{\phi_1 \max(m_i/m_{i'}, m_{i'}/m_i)}{\left(\phi_2(z_i/m_i + z_{i'}/m_{i'}) + \phi_3/R_{i,i'}\right)^2}. \tag{8.2}$$

According to the gravitation obtained in (8.2), we split the vehicles into multiple aggregation groups, which are denoted as $\{\mathcal{V}\}$. Based on this aggregation, we leverage a multi-agent learning approach to optimize edge resource allocation. Since the vehicles in the edge network have computing and communication capabilities, they can act as agents to learn the optimal edge scheduling strategies. To minimize the task offloading costs under delay constraints, the optimization problem is given in the following form:

$$\min_{\{\beta_{i,j,k}, \delta_{i,j,k,l}\}} \sum_{V_q \in \mathcal{V}} \sum_{i=1}^{|V_q|} \sum_{j=1}^{J_i} \sum_{k=1}^{|V_q|} \beta_{i,j,k} \sum_{l=1}^{L} \delta_{i,j,k,l} C_{i,j} z_i$$

$$C1: \sum_{k=1}^{V_q} \beta_{i,j,k} = 1, \quad \forall V_q \in \mathcal{V}, i, k \in V_q, j \in J_i$$

$$C2: \beta_{i,j,k} = 0, \quad \forall V_q \in \mathcal{V}, i \in V_q, j \in J, k \notin V_q \tag{8.3}$$

$$C3: \sum_{k=1}^{V_q} \beta_{i,j,k}(T_{i,j,k}^{\text{tran}} + T_{i,j,k}^{\text{exe}}) \leq T_{i,j}^{\max}, \quad \forall V_q \in \mathcal{V}, i, k \in V_q, j \in J_i$$

where constraint C1 ensures that a task can be offloaded at most to only one vehicle for processing, C2 indicates that task offloading occurs only between vehicles belonging to the same aggregation group, and C3 shows that the time consumption, including the transmission and execution time, should be within the delay constraints of the tasks. Problem (8.3) is an integer programming problem and has been proved to be NP complete.

Let $U_{V_q} = \sum_{i=1}^{|V_q|} \sum_{j=1}^{J_i} \sum_{k=1}^{|V_q|} \beta_{i,j,k} \sum_{l=1}^{L} \delta_{i,j,k,l} C_{i,j} z_i$. The target function of (8.3) can be written as $\min \sum_{V_q \in \mathcal{V}} U_{V_q}$. According to C2, there is no offloading correlation between the different aggregation groups. Thus, to address problem (8.3), we turn to minimize U_{V_q} by adopting a multi-agent deep deterministic policy gradient (MADDPG) learning approach, where $V_q \in \mathcal{V}$. The number of learning iterations is represented by the time slot t. For vehicle i belonging to aggregation group V_q, its action taken at time slot t is $a_i^t = \{\beta_{i,j,k}^t, \delta_{i,j,k,l}^t\}$, where $i, k \in V_q, j \in J_i$ and $l \in L$. Then, the action set of the multiple agents is given as $A^t = \{a_i^t\}$. The state at time slot t can be presented as $S^t = \{T_{i,j}^{\text{rem},t}, \Gamma_k^t\}$, where $T_{i,j}^{\text{rem},t}$ and Γ_k^t are the remaining completion time of the task $w_{i,j}$ and the set of tasks that have been queued for processing in vehicle k in time slot t, respectively. Taking action A^t in state S^t, the learning system of V_q gains the reward

$$Q_q^t(S^t, A^t) = \sum_{i=1}^{|V_q|} \sum_{j=1}^{J_i} \sum_{k=1}^{|V_q|} \beta_{i,j,k}^t \sum_{l=1}^{L} \delta_{i,j,k,l}^t C_{i,j} z_i. \tag{8.4}$$

The main goal of multi-agent learning in group V_q is to find the optimal action strategy for the agents to minimize the group's task offloading costs, presented as

$$Q_q(S^0, A) = \mathbb{E}\left[\sum_{t=0}^{\infty} \xi Q_q^t(S^t, A^t)|S^0\right], \tag{8.5}$$

where ξ is a discount coefficient that indicates the effect of a future reward on the current actions, and $0 < \xi < 1$.

The DTN and the multi-agent learning system operate cooperatively in scheduling the vehicular edge service. On the one hand, the DTN determines the distributed learning environments of the multiple agents by aggregating vehicular groups under the guidance of the parameters $\Phi = \{\phi_1, \phi_2, \phi_3\}$. This aggregation improves the supply and demand matching of edge resources and reduces the multi-agent learning complexity. On the other hand, the multi-agent learning results, that is, the task offloading target selection and edge resource allocation, affect the vehicular edge service performance and the performance indicators can be used in turn to evaluate the pros and cons of the aggregation mechanism, to adjust the aggregation parameter set Φ. These two parts iteratively interact and update to adapt to the changes in application scenarios.

8.2.3 Illustrative Results

We evaluate the performance of our proposed vehicular edge task offloading schemes based on real traffic data sets, which are extracted from the historical mobility traces of taxi cabs in the San Francisco Bay area. There are approximately 500 cabs, and the average time interval for their GPS coordinate updates is less than 10 seconds [112]. To investigate the influence of traffic environment characteristics on the offloading scheme performance, we further divide the Bay Area into six square areas. We consider a scenario in which the computation capacities of the vehicles are randomly taken from (10, 20) units. The computation resource requirements, data size, and maximum tolerable latency of the tasks are randomly chosen from (30, 50) units, (5, 10) MB, and (0.5, 2) seconds, respectively. In addition, there are five orthogonal channels for offloading transmissions, and the bandwidth of each channel is 0.3 MHz.

Figure 8.4 shows the offloading costs under different scheduling schemes. Compared with the other two schemes, our proposed MADDPG obtains the lowest cost. In the independent learning scheme, each vehicle works as an agent to aware edge service environments and makes self-interested offloading actions without interactions among the agents. This independent decision-making approach can create a resource surplus or shortage between some vehicular service pairs, thereby undermining the offloading efficiency of the whole system. In the MADDPG without aggregation, all the agents in the same area adopt joint decision making. Due to the complexity of the vehicle topology and potential service relations, in this scheme, it is difficult to reach the optimal offloading strategy under constrained learning iterations. In contrast to the previous two schemes, the MADDPG scheme aggregates vehicular agents based on DTN-aided edge service matching, which helps the scheme to realize low-complexity multi-agent collaborative learning under the premise of efficient resource utilization and obtains the lowest cost.

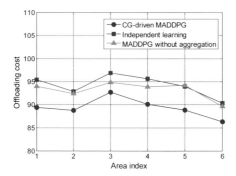

Fig. 8.4 Comparison of offloading costs with different schemes

Figure 8.5 presents the convergence of the MADDPG learning scheme. We randomly select two agents from areas 3 and 5, respectively. All the agents' learning converges around 3,300 iterations. Furthermore, this figure demonstrates that the difference in edge network characteristics and aggregation groupings between the two areas has little effect on the convergence performance.

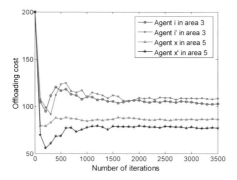

Fig. 8.5 Convergence of MADDPG learning

8.3 DT for Vehicular Edge Caching

Along with the proliferation of smart vehicles and powerful IoV applications, the huge amounts and high diversity of content need to be disseminated and shared between interactive vehicles under stringent delay constraints. However, due to limited spectrum resources, it is challenging for current wireless systems to deliver content while meeting such requirements, especially in heavy traffic scenarios with high vehicle density.

Vehicular edge caching is a promising paradigm for addressing this challenging issue. Edge caching technology locates popular content close to end users via distributed cache vehicles and RSUs and considerably accelerates the responsiveness of content acquisition from the edge, compared to fetching them from remote content providers. However, unstable communications and the highly dynamic topology between smart vehicles and RSUs still pose critical challenges in designing optimal caching schemes for vehicular edge networks. In practice, an individual edge cache server always has constrained storage space, which makes it impossible for a single server to hold multiple large files at the same time. Moreover, when the cache servers are equipped on several RSUs, the limited coverage range of an individual RSU can lead to short communication durations and small amounts of data delivered.

To effectively utilize the constrained cache and communication resources with dynamic topology, cooperative caching needs to be leveraged, where content subscribers can be served by multiple caching servers. Moreover, to make full use of the caching capabilities of smart vehicles, social interactions among the vehicles can be utilized to improve content dispatch efficiency. The social characteristics of the vehicles are basically related to their drivers, who determine their content preferences and daily driving routines and affect the other vehicles that may be encountered on the road or in parking lots.

Integrating socially aware smart vehicles and the mobile edge computing framework also requires addressing the challenges brought about by socially aware smart vehicles. For instance, vehicular social characteristics are time varying and can change dynamically according to content popularity, traffic density, and vehicle speeds. Furthermore, owing to the mobility of vehicles, highly intermittent connectivity between vehicular content providers and subscribers can seriously undermine the efficiency of socially aware content transmission. In addition, the cooperation between vehicular cache resources needs to cater to road traffic distribution, channel quality, and content popularity. Thus, supporting delay-bounded content delivery over vehicular social networks with multiple cache-enabled smart vehicles is a challenge.

DT technology can be used to address the above challenges. In socially aware vehicular edge caching networks, the DT approach can enable cache controllers to grasp the social relations between vehicles, understand the vehicle flow distribution, and effectively allocate communication and storage resources for content delivery. In this section, we propose a DT-empowered content caching mechanism for socially aware vehicular edge networks [111]. We present a DT-based vehicular edge caching framework that comprehensively captures vehicular social features and improves caching scheduling in highly dynamic vehicular networks. Moreover, by applying a deep deterministic policy gradient (DDPG) learning approach, we propose an optimal vehicular caching cloud formulation and edge caching resource arrangement that maximize the system's utility in diverse traffic environments.

8.3.1 System Model

Figure 8.6 shows a DT-empowered vehicular social edge network. We consider an intelligent transport system in urban areas, where smart vehicles provide various powerful applications, such as smart navigation, online video, and interactive gaming. The implementation of these applications always requires content generated by the data centre, which is located in the core network. The required content is classified into G types. Each type of content is described in three terms, as $T_g = \{f_g, t_g^{\max}, \mu_g\}$, and $g \in G$, where f_g is the size of content type g, $t_g^{\max}$ is its maximum delay tolerance, and μ_g is the delay sensitivity coefficient that can be taken as the utility gained from a unit time reduction compared to $t_g^{\max}$ during the content delivery process.

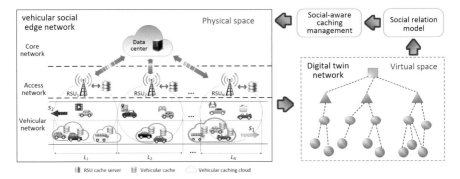

Fig. 8.6 A DT-empowered vehicular social edge network

To form access networks and provide data to vehicular content subscribers, N RSUs are located along bidirectional roads that can receive content from the data centre and then relay it to the vehicles. The diameters of the regions covered by these RSUs are $\{L_1, L_2, ..., L_N\}$, respectively. Each RSU is equipped with an edge caching server. The caching capabilities of these servers are $\{C_1, C_2, ..., C_N\}$, respectively. To avoid long transmission latencies between the data centre and the vehicles, the servers can retrieve popular content from the centre and store them in their cache for later use.

Besides being cached in RSUs, content can also be pre-stored in smart vehicles. Cache-enabled smart vehicles on the road act as content carriers and forward cached data to vehicles they encounter through V2V communication. To fully exploit V2V content delivery, vehicular social relations are leveraged in edge cache management. When the supply and demand content between vehicles is consistent and the communication link for data delivery can be established, we say that the vehicles are socially related. From this viewpoint, vehicular social relations are characterized by two elements. One element is the content matching between the supply and demand sides, and the other is the communication contact rate of the vehicles. We consider that the vehicles in this system demand G types of content with probability

$\beta = \{\beta_1, \beta_2, ..., \beta_G\}$, respectively, where $\sum_{g \in G} \beta_g \leq 1$. When a vehicle with type g content in its cache is on the road, the probability of encountering a vehicle that needs exactly this type of content is β_g. Thus, the content-matching element can be described by the probability β. The communication contact rate is defined as the number of vehicles with which a given vehicle can be associated in a unit time while it is driving.

8.3.2 DT-Empowered Content Caching

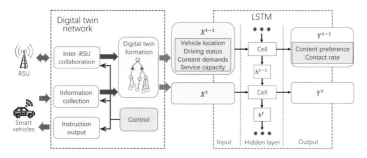

Fig. 8.7 DT and LSTM-based social model construction

Figure 8.7 illustrates the main framework of the proposed DT and a social model construction approach based on long short-term memory (LSTM). The DTN consists of five modules, where the information collection module obtains vehicular network states from smart vehicles through V2R communication. The control module determines the update cycle and adjusts the data type and interactive frequency in information collection. The adjustment will be issued to the smart vehicles through the instruction output module, thereby changing the vehicles' state sampling and reporting mode. After establishing the DT, which offers a virtual representation of the physical vehicular network, we use an LSTM recurrent network to extract the social features from the received data sets.

We use $\psi_g(\xi_g)$ to denote the accuracy of the social model that reflects the relations between the supply and demand of vehicles for type g content, where ξ_g is the amount of system information gathered by DT to train the LSTM network and obtain the social model $\{\beta, s_1, s_2\}$. The value of $\psi_g(\xi_g)$ is the modulus ratio of the estimated social model parameters to those of the true model, and $0 \leq \psi_g(\xi_g) \leq 1$. Since more information would help improve the model's accuracy, $\psi_g(\xi_g)$ is a monotonically increasing function in terms of ξ_g.

In the proposed vehicular edge caching network, to improve the delivery time efficiency while reducing transmission costs, the content needs to be efficiently prestored in appropriate cache nodes. Moreover, as the caching arrangement depends on the vehicular social model obtained from the DT-empowered LSTM system, the more information gathered by the DTN, the higher the accuracy of the model. However, the information collection process incurs a V2R communication cost. Thus, the trade-off between the V2R communication cost and model accuracy and its impact on the caching system utility also need to be considered in the cache scheduling.

Let x_g and $\mathcal{Y}_g = \{y_{g,1}, y_{g,2}, ..., y_{g,N}\}$ denote the probability of pre-storing type g content in the vehicular caching cloud and in the caching servers equipped on RSUs, respectively. The size of the content segment cached in a vehicle is Q_g. The proposed optimal edge caching problem, which maximizes the utility of the caching system under the constraints of node cache capacity and content delivery delay, can therefore be formulated as

$$
\begin{aligned}
\max_{\{x_g, \mathcal{Y}_g, Q_g, \xi_g\}} U = &\sum_{g \in G} \{ \sum_{v' \in \mathcal{V}_2} \sum_{n \in N} \Psi_g(\xi_g)\beta_g [x_g(\mu_g(t_g^{\max} \\
& -t_{g,v'}^V) - f_g\varsigma_v) + (1 - x_g)w_{g,v',n}(\mu_g(t_g^{\max} \\
& -t_{g,v',n}^R) - f_g(\varsigma_r + (1 - y_{g,n})\varsigma_c))] - \xi_g\varsigma_r \}
\end{aligned}
$$

$$
\begin{aligned}
\text{s.t.} \quad &\text{C1}: \quad 0 \leqslant x_g \leqslant 1, \quad g \in G, \\
&\text{C2}: \quad 0 \leqslant \mathcal{Y}_g \leqslant 1, \quad g \in G, \\
&\text{C3}: \quad e_v \leqslant C_v, \quad v \in \mathcal{V}_1, \\
&\text{C4}: \quad \sum_{g \in G} y_{g,n} f_g \leqslant C_n, \quad n \in N, \\
&\text{C5}: \quad \mathbf{1}\{x_g > 0\}t_{g,v'}^V \leqslant t_g^{\max}, g \in G, v' \in \mathcal{V}_2, n \in N, \\
&\text{C6}: \quad \mathbf{1}\{y_{g,n} > 0\}t_{g,v',n}^R \leqslant t_g^{\max}, g \in G, v' \in \mathcal{V}_2, n \in N, \\
&\text{C7}: \quad Q_g \leqslant Q_v^{\max}, \quad g \in G, \\
&\text{C8}: \quad \sum_{g \in G} \xi_g \leqslant \xi^{\max}, \quad \xi_g > 0, \quad g \in G,
\end{aligned}
\tag{8.6}
$$

where $\mathcal{V}_1$ and $\mathcal{V}_2$ denote the sets of the content provider and subscriber vehicles in an area, respectively; $\Psi_g(\xi_g)$ is an influence function that presents the impact of social model deviation caused by different amounts of gathered information on the system's utility; and $w_{g,v',n}$ is the probability that vehicle v' is located within the coverage of RSU n and obtains type g content from the cache server equipped on this RSU in V2R mode.

In (8.6), the first two constraints show the range of the caching probability. Constraints C3 and C4 guarantee that the amount of content on a vehicle and on an RSU server should not exceed the maximum storage capacity of the respective caching node. Constraints C5 and C6 ensure the time cost for type g content remains within its delay constraint. Constraint C7 indicates that the size of the content segment cached in a vehicle should not exceed the upper limit. The last constraint ensures that the amount of information related to type g content is positive and the total amount of gathered information should not exceed the maximum threshold $\xi^{\max}$.

In the proposed optimal caching problem, the edge cache scheduling relies on the social model built, while, in the model construction, the adjustment of information collection depends on its effect on the system's utility. Moreover, due to possible content segmentation and cache resource sharing, there exists strong correlation between the various types of content cached in heterogeneous edge caching nodes. These features make solving problem (8.6) a critical challenge. To address this issue, we propose a DDPG learning–based iterative approach. In each iteration, we first obtain the cache scheduling strategies according to a given social model and then modify the amount of information gathered in model construction based on the determined caching strategies. The iteration continues until the system's utility converges.

8.3.3 Illustrative Results

We evaluate the performance of the proposed DT-empowered and socially aware edge caching schemes based on vehicular traffic data sets gathered in different areas. We consider a scenario in which one to three RSUs are randomly located in each area. The data storage capacity of the cache server equipped on each RSU is randomly set within the interval (300, 700) MB. There are 10 types of content requirements, of which the content size, maximum delay tolerance, and delay sensitivity coefficient are randomly chosen from $(10, 100)$ MB, $(0.5, 3)$ seconds, and $(0.1, 0.3)$, respectively.

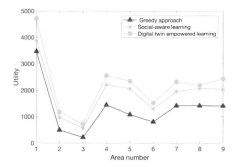

Fig. 8.8 Comparison of the caching utilities of multiple areas under different schemes

Figure 8.8 compares the utilities of multiple areas with different edge caching scheduling schemes. Our proposed DT-empowered learning approach gains the highest utilities in all the urban areas compared to the others. Here, the greedy approach, which obtains the lowest utility, arranges the content storage in the edge cache nodes only according to content popularity and ignores the social relations between smart vehicles and thus fails to make full use of the communication contacts between vehicles to implement V2V data delivery. In contrast to this approach, the socially aware learning scheme takes the content delivery among vehicles directly into account and dynamically allocates cache and communication resources based on the content requirements and known environmental characteristics, thus achieving higher utility. However, its social feature perception mode is fixed, which can increase detection costs or reduce perception accuracy. Unlike the two previous schemes, the one we proposed leverages DT to reflect the vehicular network states while adaptively adjusting social model construction strategies with balanced accuracy and costs, thus resulting in the highest caching utility.

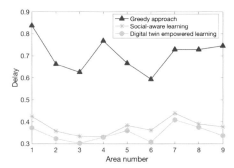

Fig. 8.9 Comparison of the content acquisition delays of multiple areas under different schemes

Figure 8.9 compares the content acquisition delays under different schemes in multiple areas. Our proposed DT-empowered learning scheme outperforms the other two approaches. Since this scheme smartly utilizes vehicular social relations and caching capacity in enabling direct data delivery between vehicles, the content acquisition delay is reduced. It is worth noting that, although in a few areas, such as area 3 in Fig. 8.8 and area 4 in Fig. 8.9, the performance of the DT-empowered learning scheme is close to that of the socially aware learning approach, in all areas as a whole, the utility (delay) of the DT-empowered scheme is increased (decreased) by 17% (10%), on average, over the simple socially aware scheme. Since both these schemes leverage vehicular social relations to schedule cache resources, the difference in their performance is smaller than the performance gap between the socially aware schemes and the greedy approach, which ignores vehicular social relation effects. Moreover, the performance gain provided by the DT mechanism is affected by the different vehicle distributions, driving states, and caching capacities in various areas. Therefore, there are differences in the gain effects of DT in these areas.

References

1. K. Främling, J. Holmström, T. Ala-Risku, M. Kärkkäinen, Product agents for handling information about physical objects, Report of Laboratory of Information Processing, Helsinki University of Technology, 2003.
2. M. Shafto, M. Conroy, R. Doyle, E. Glaessgen, C. Kemp, J. LeMoigne, and L. Wang, Modeling, simulation, information technology and processing roadmap, 2010.
3. U. S. A. Force, Global horizons final report, tech. rep. Global Science and Technology Vision, 2013.
4. M. Grieves and J. Vickers, Digital Twin: Mitigating Unpredictable, Undesirable Emergent Behavior in Complex Systems. In:M. Grieves, J. Vickers (eds.) Transdisciplinary Perspectives on Complex Systems: New Findings and Approaches, Springer. 85–113(2017)
5. F. Tao, H. Zhang, A. Liu, and A.Y. C.Nee, Digital twin in industry: State-of-the-art. IEEE Trans. Ind. Informat. 15(4), 2405–2415 (2018)
6. Gartner, Gartner's Top. 10 strategic technology trends for 2018, 2018. `https://www.gartner.com/smarterwithgartner/gartners-top-10-technology-trends-2018`
7. Gartner, Gartner's Top. 10 strategic technology trends for 2019, 2019. `https://www.gartner.com/smarterwithgartner/gartners-top-10-technology-trends-2019`
8. B. Bielefeldt, J. Hochhalter, and D. Hartl, Computationally efficient analysis of sma sensory particles embedded in complex aerostructures using a substructure approach, in Asme Conference on Smart Materials, 2015
9. Y. Lu, X. Huang, K. Zhang, S. Maharjan, and Y. Zhang, Low-latency federated learning and blockchain for edge association in digital twin empowered 6g networks. IEEE Trans. Industrial Informatics. 17(7), 5098–5107 (2020)
10. Y. Zheng, S. Yang, and H. Cheng, An application framework of digital twin and its case study. Journal of Ambient Intelligence & Humanized Computing. 10, 1141–1153(2019)
11. R. Sderberg, K. Wrmefjord, J. S. Carlson, and L. Lindkvist, Toward a digital twin for real-time geometry assurance in individualized production. Cirp Annals Manufacturing Technology. 66(1), 137–140 (2017)
12. Z. Liu, N. Meyendorf, and N. Mrad, The role of data fusion in predictive maintenance using digital twin, in AIP conference proceedings. 1949(1), (2018)
13. D. Liu, K. Guo, B. Wang, and Y. Peng, Summary and perspective survey on digital twin technology. Chinese Journal of Scientific Instrument. 39(11), 1–10 (2018)
14. G. Schroeder, C. Steinmetz, C. E. Pereira, I. Muller, N. Garcia, D. Espindola and R. Rodrigues, Visualising the digital twin using web services and augmented reality, in IEEE 14th International Conference on Industrial Informatics (INDIN), Poitiers, France, 19–21 Jul 2016 (IEEE, 2016), pp. 522–527
15. M. Milton, C. D. L. O, H. L. Ginn, A. Benigni, Controller-Embeddable Probabilistic Real-Time Digital Twins for Power Electronic Converter Diagnostics. IEEE Trans. Power Electronics. 35(9), 9850–9864 (2020)

© The Author(s) 2024

Y. Zhang, *Digital Twin*, Simula SpringerBriefs on Computing 16,
https://doi.org/10.1007/978-3-031-51819-5

16. H. Dang, M. Tatipamula, and H. Nguyen, Cloud-based Digital Twinning for Structural Health Monitoring Using Deep Learning. IEEE Trans. Industrial Informatics. 18(6), 3820–3830 (2021)

17. G. Ji, J. Hao, J. Gao, and C. Lu, Digital Twin Modeling Method for Individual Combat Quadrotor UAV, in IEEE 1st International Conference on Digital Twins and Parallel Intelligence (DTPI), Beijing, China, 15 Jul–15 Aug 2021(IEEE, 2021), pp. 1–4

18. T. Mukherjee and T. DebRoy, A digital twin for rapid qualification of 3D printed metallic components. Applied Materials Today. 14, 59–65(2019)

19. Modelica-a unified object-oriented language for systems modeling-language specification-version 3.3. [Online]. Available: https://www.modelica.org/documents/ModelicaSpec33.pdf. (2012)

20. M. W. Rohrer and I. W. McGregor, Simulating reality using automod, in Proceedings of the Winter Simulation Conference, San Diego, CA, USA, 08–11 Dec 2002(IEEE, 2002), pp. 173–181

21. W. B. Nordgren, Flexible simulation (flexsim) software: Flexsim simulation environment, in Proceedings of the 35th conference on Winter simulation: driving innovation, 2003, pp. 197–200

22. Delmia global operations. [Online]. Available: https://www.3ds.com/products-services/delmia/. (2019)

23. M. Schluse, M. Priggemeyer, L. Atorf and J. Rossmann, Experimentable Digital Twins-Streamlining Simulation-Based Systems Engineering for Industry 4.0. IEEE Trans. Industrial Informatics. 14(4), 1722–1731 (2018)

24. G. Bachelor, E. Brusa, D. Ferretto and A. Mitschke, Model-Based Design of Complex Aeronautical Systems Through Digital Twin and Thread Concepts. IEEE Systems Journal. 14(2), 1568–1579 (2020)

25. Y. Lu, S. Maharjan and Y. Zhang, Adaptive edge association for wireless digital twin networks in 6G. IEEE Internet of Things Journal. 8(22), 16219–16230 (2021)

26. T. Liu, L. Tang, W. Wang, Q. Chen and X. Zeng, Digital-Twin-Assisted Task Offloading Based on Edge Collaboration in the Digital Twin Edge Network. IEEE Internet of Things Journal. 9(2), 1427–1444 (2022)

27. W. Sun, H. Zhang, R. Wang and Y. Zhang, Reducing Offloading Latency for Digital Twin Edge Networks in 6G. IEEE Trans. Vehicular Technology. 69(10), 12240–12251 (2020)

28. Park, K. T. , Son, Y. H. , Ko, S. W. and Noh, S. D, Digital twin and reinforcement learning-based resilient production control for micro smart factory. Applied Sciences. 11(7), 2977(2021)

29. Xia. K, Sacco. C, Kirkpatrick. M, Saidy. C and Harik. R, A digital twin to train deep reinforcement learning agent for smart manufacturing plants: environment, interfaces and intelligence. Journal of Manufacturing Systems. 58, 210–230(2021)

30. X. Xu, B. Shen, S. Ding and et al, Service offloading with deep Q-network for digital twinning-empowered internet of vehicles in edge computing. IEEE Trans. Industrial Informatics. 18(2), 1414–1423 (2022)

31. G. Shen, L. Lei, Z. Li, S. Cai, L. Zhang, P. Cao and X. Liu. Deep reinforcement learning for flocking motion of multi-UAV systems: learn from a digital twin," IEEE Internet of Things Journal. 9(13), 11141–11153(2021)

32. Y. Lu, X. Huang, K. Zhang, S. Maharjan and Y. Zhang, Communication-efficient federated learning for digital twin edge networks in industrial IoT. IEEE Trans. Industrial Informatics. 17(8), 5709–5718 (2021)

33. W. Sun, S. Lei, L. Wang, Z. Liu and Y. Zhang, Adaptive Federated Learning and Digital Twin for Industrial Internet of Things. IEEE Trans. Industrial Informatics. 17(8), 5605–5614 (2021)

34. J. Zhang, Y. Liu, X. Qin and X. Xu, Energy-efficient federated learning framework for digital twin-enabled industrial internet of things, in IEEE 32nd Annual International Symposium on Personal, Indoor and Mobile Radio Communications (PIMRC), Helsinki, Finland, 13–16 Sep 2021 (IEEE 2021) pp. 1160–1166

35. W. Sun, N. Xu, L. Wang, H. Zhang and Y. Zhang, Dynamic digital twin and federated learning with incentives for air-ground networks. IEEE Trans. Network Science and Engineering. 9(1), 321–332 (2022)

36. L. Jiang, H. Zheng, H. Tian, S. Xie and Y. Zhang, Cooperative federated learning and model update verification in blockchain empowered digital twin edge networks. IEEE Internet of Things Journal. 9(13), 11154–11167 (2021)

37. Y. Lu, X. Huang, K. Zhang, S. Maharjan and Y. Zhang, Communication-efficient federated learning and permissioned blockchain for digital twin edge networks. IEEE Internet of Things Journal. 8(4), 2276–2288 (2021)

38. J. Pang, Y. Huang, Z. Xie, J. Li and Z. Cai, Collaborative city digital twin for the COVID-19 pandemic: a federated learning solution. Tsinghua Science and Technology. 26(5), 759–771 (2021)

39. Z. Zhou, Z. Jia, H. Liao, W. Lu, S. Mumtaz, M. Guizani and M. Tariq, Secure and latency-aware digital twin assisted resource scheduling for 5G edge computing-empowered distribution grids. IEEE Trans. Industrial Informatics. 18(7), 4933–4943 (2021)

40. A. Shamsi, H. Asgharnezhad, et al, An uncertainty-aware transfer learning-based framework for COVID-19 diagnosis. IEEE Trans. Neural networks and Learning Systems. 32(4), 1408–1417 (2021)

41. K. Zhang, D. Si, W. Wang, J. Cao and Y. Zhang, Transfer learning for distributed intelligence in aerial edge networks. IEEE Wireless Communications. 28(5), 74–81 (2021)

42. I. Ntinou, E. Sanchez, A. Bulat, M. Valstar and G. Tzimiropoulos, A transfer learning approach to heatmap regression for action unit intensity estimation. IEEE Trans. Affective Computing. 14(1), 436–450 (2023)

43. W. Wang, L. Tang, C. Wang and Q. Chen, Real-time analysis of multiple root causes for anomalies assisted by digital twin in NFV environment. IEEE Trans. Network and Service Management. 19(2), 905–921 (2022)

44. Y. Xu, Y. Sun, X. Liu and Y. Zheng, A digital-twin-assisted fault diagnosis using deep transfer learning. IEEE Access. 7, 19990–19999 (2019)

45. S. Liu, H. Shen, J. Li, Y. Lu and J. Bao, An adaptive evolutionary framework for the decision-making models of digital twin machining system, in IEEE 17th International Conference on Automation Science and Engineering (CASE), Lyon, France, 23–27 Aug 2021(IEEE, 2021), pp. 771-776

46. F. Zhu, Z. Li, S. Chen, and G. Xiong, Parallel Transportation Management and Control System and Its Applications in Building Smart Cities. IEEE Trans. Intell. Transp. 17(6), 1576–1585 (2016)

47. H. Dong, B. Ning, Y. Chen, X. Sun, D. Wen, Y. Hu, and R. Ouyang, Emergency Management of Urban Rail Transportation Based on Parallel Systems," IEEE Trans. Intell. Transp. 14(2), 627–636 (2013)

48. S. Zhou, Y. Sun, Z. Jiang, Z. Niu, Exploiting Moving Intelligence: Delay-Optimized Computation Offloading in Vehicular Fog Networks. IEEE Commun. Mag. 57(5), 49–55 (2019)

49. S. Haag and R. Anderl, Digital Twin–Proof of Concept. Manuf. Lett. 15, 64–66 (2018)

50. B. Minerva, G. Lee, and N. Crespi, Digital Twin in the IoT Context: A Survey on Technical Features, Scenarios, and Architectural Models. P. IEEE. 108(10), 1785–1824 (2020)

51. C. Gehrmann, and M. Gunnarsson, A Digital Twin Based Industrial Automation and Control System Security Architecture IEEE Trans. Ind. Inform. 16(1), 669–680 (2020)

52. P. Jia, X, Wang, and X. Shen, Digital-Twin-Enabled Intelligent Distributed Clock Synchronization in Industrial IoT Systems. IEEE Internet Things J.8(6), 4548–4559 (2020)

53. L. Jiang, S. Xie, S. Maharjan, and Y. Zhang, Joint Transaction Relaying and Block Verification Optimization for Blockchain Empowered D2D Communication. IEEE Trans. Veh. Technol. 69(1), 828–841 (2020)

54. X. Huang, S. Leng, S. Maharjan and Y. Zhang, Multi-Agent Deep Reinforcement Learning for Computation Offloading and Interference Coordination in Small Cell Networks. IEEE Trans. Veh. Tech. 70(9), 9282–9293 (2021)

55. Y. Dai, Y. Guan, K. Leng, and Y. Zhang, Reconfigurable Intelligent Surface for Low-latency Edge Computing in 6G. IEEE Wirel. Commun.. 28(6), 72–79 (2021)

56. B. Li, Z. Fei, and Y. Zhang, UAV Communications for 5G and Beyond: Recent Advances and Future Trends. IEEE Internet Things J.. 6(2), 2241–2263 (2019)

57. L. Jiang, B. Chen, S. Xie, S. Maharjan and Y. Zhang, Incentivizing Resource Cooperation for Blockchain Empowered Wireless Power Transfer in UAV Networks. IEEE Trans. Veh. Tech. 69(12), 15828–15841 (2020)

58. N. Abbas, Y. Zhang, A. Taherkordi, and T. Skeie, Mobile edge computing: A survey. IEEE Internet of Things Journal. 5(1), 450–465 (2017)

59. K. Zhang, Y. Zhu, S. Maharjan, and Y. Zhang, Edge intelligence and blockchain empowered 5g beyond for the industrial internet of things. IEEE Network. 33(5), 12–19 (2019)

60. Y. Dai, D. Xu, S. Maharjan, G. Qiao, and Y. Zhang, Artificial intelligence empowered edge computing and caching for internet of vehicles. IEEE Wireless Communications. 26(3), 12–18 (2019)

61. J. Kang, R. Yu, X. Huang, M. Wu, S. Maharjan, S. Xie and Y. Zhang, Blockchain for Secure and Efficient Data Sharing in Vehicular Edge Computing and Networks. IEEE Internet of Things J. 6(3), 4660–4670 (2019)

62. Z. Li, J. Kang, R. Yu, D. Ye, Q. Deng and Y. Zhang, Consortium Blockchain for Secure Energy Trading in Industrial Internet of Things. IEEE Trans. Ind. Inform. 14(8), 3690–3700 (2018)

63. J. Xie, K. Zhang, Y. Lu, and Y. Zhang, Resource-Efficient DAG Blockchain with Sharding for 6G Networks. IEEE Netw. 36(1), 189–196 (2022)

64. H. Liu, Y. Zhan, and T. Yang, Blockchain-Enabled Security in Electric Vehicles Cloud and Edge Computing. IEEE Netw. 32(3), 78–83 (2018)

65. M. Xu, W. Ng, W. Lim, J. Kang, Z. Xiong, D. Niyato, Q. Yang, X. Shen, and C. Miao, A Full Dive into Realizing the Edge-Enabled Metaverse: Visions, Enabling Technologies, and Challenges. IEEE Communications Surveys & Tutorials. 25(1), 656–700 (2022)

66. W. Wang, D. Hoang, P. Hu, Z. Xiong, D. Niyato, P. Wang, and Y. Wen, A Survey on Consensus Mechanisms and Mining Strategy Management in Blockchain Networks. IEEE Access. 7, 22328–22370 (2019)

67. W. Saad, M. Bennis, and M. Chen, A vision of 6g wireless systems: Applications, trends, technologies, and open research problems. IEEE network. 34(3), 134–142 (2019)

68. X. You, C.-X. Wang, J. Huang, X. Gao, Z. Zhang, M. Wang, Y. Huang, C. Zhang, Y. Jiang, J. Wang et al., Towards 6g wireless communication networks: Vision, enabling technologies, and new paradigm shifts. Science China Information Sciences. 64(1), 1–74 (2021)

69. W. Jiang, B. Han, M. A. Habibi, and H. D. Schotten, The road towards 6g: A comprehensive survey. IEEE Open Journal of the Communications Society. 2, 334–366 (2021)

70. L. U. Khan, W. Saad, D. Niyato, Z. Han, and C. S. Hong, Digital-twin-enabled 6g: Vision, architectural trends, and future directions. IEEE Communications Magazine. 60(1), 74–80 (2022)

71. H. X. Nguyen, R. Trestian, D. To, and M. Tatipamula, Digital twin for 5g and beyond. IEEE Communications Magazine. 59(2), 10–15 (2021)

72. K. Zhang, J. Cao, and Y. Zhang, Adaptive digital twin and multiagent deep reinforcement learning for vehicular edge computing and networks. IEEE Transactions on Industrial Informatics. 18(2), 1405–1413 (2022)

73. V. Mnih, A. P. Badia, M. Mirza, A. Graves, T. Lillicrap, T. Harley, D. Silver, and K. Kavukcuoglu, Asynchronous methods for deep reinforcement learning, in International conference on machine learning, PMLR (2016), pp. 1928–1937

74. Y. Dai, K. Zhang, S. Maharjan, and Y. Zhang, Deep reinforcement learning for stochastic computation offloading in digital twin networks. IEEE Transactions on Industrial Informatics. 17(7), 4968–4977(2021)

75. M. J. Neely, Stochastic network optimization with application to communication and queueing systems. Synthesis Lectures on Communication Networks. 3(1), 1–211(2010)

76. Y. Li, L. Lei, and M. Yan, Mobile user location prediction based on user classification and markov model, in 2019 International Joint Conference on Information, Media and Engineering (IJCIME), Osaka, Japan, 17-19 Dec 2019. (IEEE, 2019), pp. 440–444

77. W. C. Ao and K. Psounis, Approximation algorithms for online user association in multi-tier multi-cell mobile networks. IEEE/ACM Trans. Netw. 25(4), 2361–2374 (2017)

78. Y. Shi, Y. Cao, J. Liu and N. Kato, A Cross-Domain SDN Architecture for Multi-Layered Space-Terrestrial Integrated Networks. IEEE Network. 33(1), 29–35(2019)
79. P. Zhang, C. Wang, N. Kumar and L. Liu, Space-Air-Ground Integrated Multi-Domain Network Resource Orchestration Based on Virtual Network Architecture: A DRL Method. IEEE Transactions on Intelligent Transportation Systems. 23(3), 2798–2808 (2022)
80. J. Deng, Q. Zheng, G. Liu, J. Bai, K. Tian, C. Sun, Y. Yan and Y. Liu, A digital twin approach for self-optimization of mobile networks, in 2021 IEEE Wireless Communications and Networking Conference Workshops (WCNCW), Nanjing, China, 29–29 Mar 2021 (IEEE, 2021), pp. 1–6
81. W. Zhuang, Q. Ye, F. Lyu, N. Cheng and J. Ren, SDN/NFV-Empowered Future IoV with Enhanced Communication, Computing, and Caching. Proceedings of the IEEE. 108(2), 274–291 (2020)
82. M. Li, P. Si and Y. Zhang, Delay-Tolerant Data Traffic to Software-Defined Vehicular Networks with Mobile Edge Computing in Smart City. IEEE Transactions on Vehicular Technology. 67(10), 9073–9086 (2018)
83. Y. He, N. Zhao and H. Yin, Integrated Networking, Caching, and Computing for Connected Vehicles: A Deep Reinforcement Learning Approach. IEEE Transactions on Vehicular Technology. 67(1), 44–55 (2018)
84. T. Do-Duy, D. Van Huynh, O. A. Dobre, B. Canberk and T. Q. Duong, Digital Twin-aided Intelligent Offloading with Edge Selection in Mobile Edge Computing. IEEE Wireless Communications Letters. 11(4), 806–810 (2022)
85. G. Qu, H. Wu, R. Li and P. Jiao: Dmro, A deep meta reinforcement learning-based task offloading framework for edge-cloud computing. IEEE Transactions on Network and Service Management. 18(3), 3448–3459 (2021)
86. W. Sun, P. Wang, N. Xu, G. Wang and Y. Zhang, Dynamic Digital Twin and Distributed Incentives for Resource Allocation in Aerial-assisted Internet of Vehicles. IEEE Internet of Things Journal. 9(8), 5839–5852 (2022)
87. H. Zhou, T. Wu, H. Zhang and J. Wu, Incentive-Driven Deep Reinforcement Learning for Content Caching and D2D Offloading. IEEE Journal on Selected Areas in Communications. 39(8), 2445–2460 (2021)
88. X. Lin, J. Wu, J. Li, W. Yang and M. Guizani, Stochastic Digital-Twin Service Demand with Edge Response: An Incentive-Based Congestion Control Approach. IEEE Transactions on Mobile Computing. 22(4), 2402–2416 (2021)
89. S. Han, S. Xu, W. Meng and C. Li, Dense-Device-Enabled Cooperative Networks for Efficient and Secure Transmission. IEEE Network. 32(2), 100–06 (2018)
90. Y. S. Nasir and D. Guo, Multi-Agent Deep Reinforcement Learning for Dynamic Power Allocation in Wireless Networks. IEEE Journal on Selected Areas in Communications. 37(10), 2239–2250 (2019)
91. C. Wang, C. Liang, F. R. Yu and Q. Chen, L. Tang, Computation Offloading and Resource Allocation in Wireless Cellular Networks with Mobile Edge Computing. IEEE Transactions on Wireless Communications. 16(8), 4924–4938 (2017)
92. C. Liang and F. R. Yu, Distributed resource allocation in virtualized wireless cellular networks based on ADMM, in IEEE Conference on Computer Communications Workshops, Hong Kong, China, 26 Apr–01 May 2015, pp. 360–365
93. Z. Zheng, L. Song, Z. Han, G. Y. Li and H. V. Poor, A Stackelberg Game Approach to Large-Scale Edge Caching ,in IEEE Global Communications Conference(GLOBECOM), Abu Dhabi, United Arab Emirates, 09–13 Dec 2018 (IEEE, 2018), pp. 1–6
94. N. H. Tran, W. Bao, A. Zomaya, M. N. H. Nguyen and C. S. Hong, Federated Learning over Wireless Networks: Optimization Model Design and Analysis,in IEEE INFOCOM 2019 - IEEE Conference on Computer Communications. Paris, France, 29 Apr–02 May 2019 (IEEE,2019), pp. 1387-1395
95. M. Krouka, A. Elgabli, C. B. Issaid and M. Bennis, Communication-Efficient and Federated Multi-Agent Reinforcement Learning. IEEE Transactions on Cognitive Communications and Networking. 8(1), 311–320 (2022)

96. W. Zhang, D. Yang, W. Wu, H. Peng, N. Zhang, H Zhang and S. Shen, Optimizing Federated Learning in Distributed Industrial IoT: A Multi-Agent Approach. IEEE Journal on Selected Areas in Communications. 39(12), 3688–3703 (2021)

97. S. Li, G. Yang and J. Zhang, A collaborative learning framework via federated meta-learning, in IEEE 40th International Conference on Distributed Computing Systems (ICDCS),Singapore, Singapore, 29 Nov–01 Dec 2020 (IEEE,2020), pp. 289–299

98. Wei. Ybl, Z. Xiong, C. Miao, D. Niyato and H. V. Poor, Hierarchical Incentive Mechanism Design for Federated Machine Learning in Mobile Networks. IEEE Internet of Things Journal. 7(10), 9575–9588 (2020)

99. Shashi Raj. Pandey, Nguyen H. Tran, Mehdi. Bennis, Yan Kyaw. Tun, Aunas. Manzoor and Choong Seon. Hong, A Crowdsourcing Framework for On-Device Federated Learning. IEEE Transactions on Wireless Communications. 19(15), 3241–3256(2020)

100. Q. Yang, Y. Liu, T. Chen and Y. Tong, Federated Machine Learning: Concept and Applications. ACM Transactions on Intelligent Systems and Technology. 10(2), 1–19 (2019)

101. W. Y. B. Lim, J. Huang, Z. Xiong, J. Kang, D. Niyato, X. Hua, C. Leung and C. Miao, Towards Federated Learning in UAV-Enabled Internet of Vehicles: A Multi-Dimensional Contract-Matching Approach. IEEE Transactions on Intelligent Transportation Systems. 22(8), 5140–5154 (2021)

102. W. Sun, H. Zhang, R. Wang and Y. Zhang, Double auction-based resource allocation for mobile edge computing in industrial internet of things. IEEE Transaction on Industrial Informatics. 14(10), 4692–4701 (2018)

103. N. B. Shah and D. Zhou: Double or nothing, Multiplicative incentive mechanisms for crowdsourcing. The Journal of Machine Learning Research. 17(1), 5725–5776 (2016)

104. N. Cheng, W. Quan, W. Shi, H. Wu, Q. Ye, H. Zhou, W. Zhuang, X. Shen and B. Bai, A Comprehensive Simulation Platform for Space-Air-Ground Integrated Network. IEEE Wireless Communications. 27(1), 178–185 (2020)

105. Q. Song, S. Lei, W. Sun and Y. Zhang, Adaptive Federated Learning for Digital Twin Driven Industrial Internet of Things. 2021 IEEE Wireless Communications and Networking Conference (WCNC),Nanjing, China, 29 Mar–01 Apr 2021 (IEEE,2021), pp. 1–6

106. Y. Hui, X. Ma, Z. Su, N. Cheng, Z. Yin, T. H. Luan and Y. Chen, Collaboration as a service: digital twins enabled collaborative and distributed autonomous driving. IEEE Internet of Things Journal. 9(19), 18607–18619 (2022)

107. J. Liu, Y. Dong, Y. Liu, P. Li, S. Liu and T. Wang, Prediction study of the heavy vehicle driving state based on digital twin model,in IEEE International Conference on Power Electronics, Computer Applications (ICPECA), Shenyang, China, 22-24 Jan 2021 (IEEE,2021), pp. 789–797

108. T. Liu, L. Tang, W. Wang, X. He, Q. Chen, X. Zeng and H. Jiang, Resource allocation in DT-assisted internet of vehicles via edge intelligent cooperation. IEEE Internet of Things Journal. 9(19), 17608–17626 (2022)

109. X. Liao, Z. Wang, X. Zhao, K. Han, P. Tiwari, M. J. Barth and G. Wu, Cooperative ramp merging design and field implementation: a digital twin approach based on vehicle-to-cloud communication. IEEE Trans. Intelligent Transportation Systems. 23(5), 4490–4500 (2021)

110. Z. Lv, Y. Li, H. Feng and H. Lv, Deep learning for security in digital twins of cooperative intelligent transportation systems. IEEE Trans. Intelligent Transportation Systems. 23(9), 16666–16675 (2021)

111. K. Zhang, J. Cao, S. Maharjan and Y. Zhang, Digital Twin Empowered Content Caching in Social-Aware Vehicular Edge Networks. IEEE Trans. Computational Social Systems. 9(1), 239–251 (2022)

112. http://crawdad.org/epfl/mobility/20090224.

Printed in the United States
by Baker & Taylor Publisher Services